LAURIE LACEY

NIMBUS
PUBLISHING LTD

Nimbus Publishing Limited
P.O. Box 9301, Station A
Halifax, NS B3K 5N5
(902)455-4286

Laurie Lacey
e-mail: llacey@isisnet.com
the web: http://fox.nstn.ca/~trogers/plants/
Design: Kathy Kaulbach
Printed and bound in Canada

Canadian Cataloguing in Publication Data
Lacey, Laurie.
Black spirit
ISBN 1-55109-152-6
1. Crows. I. Title.
QL696.P2367L32 598.8'64 C96-950050-5

This book is dedicated to my mother,
who truly appreciates crows,
and to my late brother, Leon.

One crow sorrow, two crows joy;
three crows a letter, four crows a boy;
five crows silver, six crows gold;
seven crows a story yet to be told.

CONTENTS

Acknowledgements

I would like to extend thanks to Florence Hubley, Sheila Wilson, Linda Goodin, John McKay, the late Burnice Gilchrist, Alison Yoshioka, Phyllis Gillis, Marilyn Burns, F. Weatherbee, Doris Phillips, Iola Stronach, Elizabeth Turner, Florence Langille, Stephen Czapalay, Charles and June Maginley, Muriel Tucker, Dorothy Knowles, Norman Deale, Myrna Wilson, Mary Patterson, Joan Stiles, and Madeline Way, all of whom contributed crow stories or general information about crows. Also, thanks to Peggi Thayer for contributing to the section on crow stories and for helping to nurture my love for crows. I wish to acknowledge the contribution of a special person in my life, Penny Dexter, who did the initial editing of the manuscript, made useful suggestions for the text and cover, and gave much encouragement. Thanks are due Glenn and Evangeline Lacey and my nephew, Jon David, for taking care of Spirit during the initial period following his accident and for bringing him into my life. I wish to thank Paula Sarson, my editor at Nimbus, for her sensitive suggestions and comments, and Dorothy Blythe, managing editor at Nimbus, for her encouragement, understanding, and faith that I could pull this book together. I am grateful to my friend Earl Pentz for suggesting "Black Spirit" as the title. Finally, I wish to acknowledge my late crow friend, Spirit—this is as much his book as it is mine.

INTRODUCTION

I have been fascinated with crows and other two-winged beings for as long as I can remember. In my childhood, I would often imagine myself flying with the freedom of a bird, gliding like a crow over the tops of trees or soaring like an eagle. As an adult, I tended to neglect those wonderful "flights" of the imagination, until a crow so charmed me that my soul soared again in the world of the flying creatures.

This book contains excerpts from a diary I kept during five years that I lived with a crow and chronicles many of the activities, experiences, and adventures we shared. The book also contains anecdotes from people who have had tame crows or those who were simply fortunate enough to witness crows' intelligence at work. It also holds my own observations of crows at large, in their natural environment. I have discovered that crows are among the most intelligent birds, but they are a misunderstood population, seeming only to wait, watch, and devour our garbage. They continue to thrive, despite periodic efforts to control their numbers, while we methodically pollute the earth on which we both depend.

During the five years with my crow companion, I became increasingly aware of and curious about crows. I was able to learn much about their language and culture. As my friendship with one crow grew, we developed a mutual trust and a surprisingly clear system of communication. I recognized then the time was right to begin a diary on crow behaviour, to record information one would normally not see by casually observing crows in the wilderness. Initially, my notes were scribbled on bits of paper or loosely written in a notebook, which I often carried with me when observing crows, something I did with increasing enthusiasm. My particular crow friend, whom I eventually named "Spirit," (others knew him as "Black Jack" or "Black Spirit") was an inspiration and a stimulus, urging me to truly discover the nature of crows in general. Of course, I was especially interested in writing about my relationship with Spirit, particularly since I saw an opportunity to foster a better understanding of the crow species. In today's world where so many people lack respect for life, especially animal life, I felt it was important to share my experiences with as many people as possible, to show that animals, too, have meaningful lives as sentient creatures that warrant our respect.

When I review the pages of my diary now, I am still amazed and delighted at the ways Spirit revealed his crow nature to me. He expressed anger, he displayed a sense of humour, and he showed kindness, love, and an acceptance of my world. Just as importantly, Spirit was an example of a crow's marvellous ability to adjust to new situations and a new environment.

I have stated that *Black Spirit* was written to promote respect for and understanding of crows. While that is certainly utmost in my mind, on another level, I intend the book to serve a teaching and healing function; I hope people will read it in a beautiful natural setting, such as a hardwood forest or while relaxing on the shore of a secluded cove; I envision it being read in the flickering light of a campfire, with puffs of wood smoke swirling about and scenting its pages. And I imagine a traditional "talking circle," one person tightly gripping the "talking stick" while reading about the crow as a trickster figure. Whatever the situation or circumstance, I hope the book will be good medicine, contributing to joy-inspired living as well as stimulating interest in the observation and protection of all wildlife.

II

The crow is a clever bird with a well-developed personality. This, combined with the blackness of its feathers, has made it a common character in legend, folktale, and myth. Generally, corvids, members of the passerine bird family, are well represented in the folklore of many cultures throughout the world. According to Christopher Leahy's *The Birdwatcher's Companion*, the gray jay, or wiskedjak, for example—a cousin of the crow—is a trickster in the traditions of certain northeastern North American First Nations (p. 164).

Another cousin of the crow's, the raven, has achieved godlike status in parts of the world. It was the raven who sustained Elijah, the biblical medicine man and prophet, while he dwelt by the brook Cherith in 1 Kings 17:2-6. In the legends of Native peoples of North America, the crow and the raven are often represented as trickster figures(Leahy, p. 5; Angell, p. 51). In fact, raven stories of the Northwest Pacific Coast peoples become crow stories in the Yukon. The trickster in legend and myth is somewhat similar to culture heroes, such as Glooscap, who is the central figure in many Mi'kmaq legends; however, there are important differences between the trickster and culture heroes. Annemarie de Waal Malefijt remarks in *Religion and Culture* that the trickster is a semi-divine being, but unlike Glooscap, it is not ancestral nor particularly concerned with the welfare of human beings (p. 162). As well, tricksters are ambivalent characters, mediating between opposites, such as good/bad, sky/earth, wild/cultivated, and male/female. This ambivalence, combined with the power of transformation, make the trickster a rather versatile being, able to play many roles and to affect humans in almost every conceivable way.

According to legend, Crow was one such versatile figure. He made the world, gave it light, and made people. In native legends from the Yukon, we learn that Crow was anxious to create the world but was having problems because the heavenly bodies (sun, moon, and stars) were guarded and under the control of a powerful Chief. Crow, however, was cunning and soon was able to trick the Chief's daughter into "giving birth to him." As her son, he was able to steal the heavenly bodies from his grandfather. Then he quickly left the domain of the powerful Chief, taking these treasures with him in a box.

Crow did a lot of travelling and eventually reached a place where fox, wolf, wolverine, mink, and rabbit were fishing in a river. At that time the world was dark, and animals were able to talk just as people do now.

Crow asked the animals for fish, but they completely ignored him. He was persistent and threatened to bring "daylight" if they refused to give the fish. This threat only made the animals laugh. Yet the laughter was brief, for Crow began to open the box, allowing a single ray of light to escape. He continued until, little by little, the light escaped completely. The animals were scared, scattering to hide in the bushes, where they turned into their present forms. Crow then commanded the sun, moon, stars, and daylight to go to the sky so that no one person would ever own them again; they would be available to everyone.

Upon bringing light to the world, Crow realized sea lion owned all the existing land on the earth. This land was an island, surrounded by water, which covered the rest of the world. Crow wanted land too. So, he made a plan. He stole sea lion's son. When sea lion demanded to have his son back, Crow asked for sand or beach in exchange for the child. Sea lion consented and gave him the sand, which Crow scattered in the ocean. This sand became the land masses of the earth.

After completing the task of making the earth, Crow spent much time walking and flying about the world. This made him very lonely. So, taking poplar bark, Crow made a carving of a person, and breathed life into it, saying "Live!" He made crow and wolf people, too. But they were very shy—crow man and woman were too shy to talk to each other, as were wolf man and woman. Crow realized this was a terrible situation and that he must change things. He had crow man sit with wolf woman and wolf man sit with crow woman. As a result, crow married wolf, and wolf married crow. "That's how the world began."[1] This union symbolizes the establishment of kinship rules and the beginning of the world in the sense of man and woman living together and populating the earth.

Stories among Yukon Native peoples form part of a greater cycle of legends, explaining the heroic and creative deeds of Crow and the many adventures that assist people in understanding the world as it now exists. The stories make wonderful reading for children and adults, stirring the imagination to ponder the mysteries of life. In fact, having a basic knowledge of crow legends helped me forge a closer relationship with Spirit. The legends made me aware of the recognition crows are accorded in traditional cultures. They alerted me to the value of recognizing the clever, resourceful nature of animals and of using such recognition to teach cultural beliefs and social values. I realized what a wonderful opportunity had presented itself when I met Spirit—the trickster, the creator of the world, was standing in front of me in the form of a small black bird! I would be a first-hand witness to his intelligent, cunning nature. Perhaps in some way he would bring the crow legends to life for me. This expectation brought me closer to Spirit; because of him, I was eager to enjoy our time together, to discover new things about all crows, and to understand them better.

1 *The foregoing crow legends are summarized accounts of legends by Elders Angela Sidney, Kitty Smith, and Rachel Dawson in* My Stories are my Wealth, *1–3.*

1

ABOUT CROWS

Arthur Bent remarked in *Life Histories of North American Jays, Crows, and Titmice*, "It has been aptly stated that if a person knows only three kinds of birds one of them will be the crow" (p. 226). This degree of commonness is surely what P. A. Taverner also had in mind when he described the crow as "a large, all black bird," and left it at that. It is a pointed definition, one which is somewhat upsetting to a crow lover such as myself because it does not reflect the depth of the crow's nature. Of course it's true, most people are quite familiar with the general physical characteristics of the crow. Even their finer physical features are familiar to many earnest bird-watchers. This is partly because of their range. They are widely distributed throughout the world, with the exception of South America, New Zealand, and a number of oceanic islands. While other species are much more localized in North America, the common crow occurs continent-wide. Yet for me, the study of the physical characteristics of crows, their behaviour, and how they interact with the environment is a process of constant discovery.

My first discovery was that the term "crow" refers to approximately thirty members of the family Corvidae, of the order Passeriformes. The common crow is Corvus brachyrhynchos. The name probably comes directly from the crow "call," although nowadays "crowing" suggests the sound of a rooster rather than a crow. According to Leahy, the term is used collectively by scientists to refer to jackdaws, jays, magpies, rooks, ravens, crows, and other members of the family (p. 164).

I next discovered (brace yourself) that crows and other corvids are classified as songbirds! Why, you might ask, would a crow ever be classified as a songbird? The answer is straightforward. The crow has a brilliant bird brain and highly developed vocal muscles, giving it the ability and distinction of a songbird. Also, with the exception of species with inordinately long tail feathers, they are among the largest songbirds in the world. In fact, Leahy asserts that the Common Raven, with its large wing span, can lay claim to the title. The calls of crows and ravens are rather harsh and guttural for songbirds, so they are often dismissed by admirers of more exotic birds. However, crows are also amazing for the large vocabulary of sounds in their repertoire. They have warning calls, sounds signifying readiness for flight, and among others, an "inviting" call to other members of the group requesting their company when food is present.

With a story of crows communicating about food, Muriel Tucker from Truro, Nova Scotia, wrote in a letter to me:

I started putting out scraps, and no sooner was I back in the house than they came to investigate. They are timid birds and waited in a nearby tree, until they were sure there were no cats or strangers around. They would not come while I was still outside. Once they were sure there was no danger, they came on the fence and began calling all their relatives and pals.

Sometimes I would go outdoors and there would be no crows in sight—not one anywhere. Well, in less time than it takes to type this sentence they appeared from nowhere, asking for hand-outs. I hated to disappoint them. When they would sit on a nearby branch I would talk

to them, and they would watch me, nodding their heads.

Crows have a pecking order and are very family conscious. As we got to know each other better, they would come to sit on the back fence and watch my windows, hoping I would notice them and bring out grub, which of course, I usually did.

As far as food goes, crows are not especially fussy. They are scavengers, willing to eat virtually all animal and vegetable foods, except green plants. Insects, shellfish, and animal carcasses are mainstays of their diet. As well, crows have a habit of foraging through garbage, eating the discards of humans. A crow is able to eat almost anything because its stomach contains powerful enzymes and acids, which render harmless almost any offal, including food tainted with botulism. In fact, to a crow, a garbage bag is like a flea market table; several garbage bags are like a giant flea market on a Sunday afternoon! We know how crazy humans can get searching out bargains. Well, crows act the same way around garbage bags. So, the next time you see crows pecking through the strewn contents of your garbage bags, try to appreciate the impulse and be a little understanding.

Part of the reason for the usual appearance of more than one crow at your garbage is that the crow is a very social bird. Candace Savage wrote in *Bird Brains* that every corvid has a song and that social groups of American crows, for example, have their own particular set of song elements. She remarks that "... each social group uses a particular set of elements in its song—some the same as its neighbours, some different—which members share through imitation. If two members ... are particularly attached ... they tune their songs even more closely to each other's" (p. 86).

Certain species nest in colonies, forming huge winter roosts. Tony Angell remarked in *Ravens, Crows, Magpies and Jays* that there was a huge roost in Kansas that had over five million crows (p. 64). He doesn't specify the precise location, nor what it was like, but this is certainly an extraordinary number; I have difficulty imagining a roost of five thousand crows, much less five million! There are numerous other reports of roosts of crows numbering in the thousands.

Because crows can descend in such damaging numbers, their fondness for corn and other cash crops has resulted in increasing animosity from farmers and agricultural authorities. In the past, crop owners have used extreme measures to limit crow populations. A good example of such extremes occurred in the state of Illinois, where a crow's roost was planted "with a thousand shrapnel grenades," while the crows were away foraging. When they returned to the roost, according to Angell, the grenades were detonated, "... and at daybreak one hundred thousand crows were on the ground. It was argued that such slaughter was necessary to protect crops, ... livestock and game animals. Evidence to the contrary was ignored, as was the suggestion that corvids ... might be beneficial to human enterprise" (p. 53). In fact, crows are beneficial in cleaning up waste matter from farms and the countryside. As well, people generally overlook the contribution crows make in controlling the insect population.

In personal correspondence, Charles and June

Maginley of Mahone Bay, Nova Scotia, mention a large roost of crows, which occurred for several winters in Sydney, Nova Scotia:

Crows! When we lived in Sydney, I was fascinated by their behaviour. Each winter all the crows in Sydney would congregate at night in a roosting area where there were large trees. This area changed each year. . . .

At dawn all the crows would take off from the roosting area, and great clouds of them could be seen crossing the harbour. As evening approached they began to gather on the ice (Sydney Harbour is frozen over from January to April). We had a good view of them from our house. They would settle in two or three groups of hundreds of birds, thousands in all. The groups were always elongated ovals, not round. They did not appear to be making much noise. Some birds would take off, others would join, and a few would go from one group to another, but each patch of birds remained about the same size. A channel of open water or thin, recently frozen ice, made by ships, went up the harbour, and the crows were near it but not right at the edge. At sunset they would all take off and fly to the roosting area.

Those roosts, with the accompanying "flocking" behaviour are not solely for social reasons; they also afford protection against predators, such as the Snowy Owl or Red Tailed Hawk, and they maximize the discovery and exploitation of food sources. Also, flocking tends to reinforce the altruistic qualities of the crow. Corvids in general are known for acts of altruism towards their own kind. There are local stories and legends of crows supporting, defending, and coming to the rescue of other members of the flock who have gotten themselves into precarious situations. They are also very protective of the baby crows.

Charles and June Maginley concluded their letter by noting:

In an earlier house, a pair of crows nested nearby and eventually the young one got out of the nest and was being fed on the ground. It could manage short hops. (This period of vulnerability is probably the only thing that prevents the world from being overrun by crows.) The parents and last year's fledgling guarded it, making a big row all the time. We had a small black kitten, which was prowling in the yard—not actually hunting the chick—but the parents started to swoop down on it. The sight of that black kitten streaking for the safety of the house with a crow zooming a foot above it was something to remember! Eventually the young crow made a big hop across the road and the commotion went over there.

I haven't had the opportunity to witness crow parents protecting their young in this manner, but I have seen crows come to the rescue or defence of a comrade, who was having a flying confrontation with a hawk. The combatants were flying almost parallel, making passes at each other, when a second crow appeared on the scene, swooping down at the hawk. The latter, outnumbered by then, quickly sought the refuge and protection of a maple tree nearby.

Although crows gather in roosts for social interaction, they nest elsewhere. Crows and ravens generally make large nests in trees; although ravens are known to use rock ledges or abandoned buildings in some instances. Most crows construct their nests with sticks and mud, lining it with all sorts of materials, including strips of bark, wool, and hair. They usually lay from three to six eggs in the spring of the year. Ranging in colour from bluish green to greenish white, the eggs may be mottled, speckled, or blotched with brown or gray. Both sexes may incubate the eggs, although it is primarily the female who does this. As well, it is likely that both parents feed the hatchlings. After approximately eighteen days, the young birds emerge from their shells, and it will be roughly another thirty-five days before they acquire the feathers necessary for flight.

The crow is highly intelligent and, like other birds, has the ability to avoid humans or other creatures who may do them harm. But not only are crows astute enough to avoid such danger, there are also recorded instances of crows and ravens using "weapons" to protect their young! For example, ravens have been known to drop stones or rocks on the heads of predators to defend their nests. Crows have also been known to drop shellfish on rocks or the hard surface of roads to break them open. This is reminiscent of the fable about the thirsty crow who came upon a half-filled pitcher of water. Unable to reach the water with his beak, he collected small stones and dropped them in the pitcher until the water reached the required level.

Tony Angell in *Ravens, Crows, Magpies and Jays* tells a remarkable story about a carrion crow of northern Europe, who apparently was able to employ human tools to secure a meal (p. 81). The crow had watched men ice fishing, hauling lines with fish, rebaiting hooks, and waiting for flags to "snap-up," signalling a "strike." On one occasion when a strike occurred, the crow flew down, secured the line in its beak, and pulled at it while retreating from the hole. Before long it dropped the line, and walked back to the opening in the ice. The crow kept its weight on the line to prevent it from slipping back into the water. Taking the line in its beak again, the crow retreated until the fish was brought out onto the ice, where it was devoured by the deserving crow. Such cognitive thinking is seldom attributed to creatures other than humans.

Some ornithologists believe the corvids are the most highly evolved family of birds. Robert Powell remarks that "Corvids have the largest cerebral hemispheres, relative to their body size, of all birds." Irene M. Pepperberg published the assessment in a 1991 paper in *Cognitive Ethology* that corvids are indeed marvellously intelligent and in league with mynas and parrots. This intelligence can be readily observed from crows in the wilderness; living with one offers an even greater appreciation. They are capable of adjusting well to new circumstances or environment changes; they can learn to express themselves vocally or otherwise to humans; and they can become rather attached to "the hand" that feeds them. What five years with a clever, expressive bird has shown me of the nature of crows and the affinity among all living things is what the following pages are about.

2

THE BEGINNING

Perhaps this crow had been trying to master the intricacies of flight, or enjoying the wind surging past his wings, or perhaps he was pecking at a carcass in the middle of the highway, as crows will do when the opportunity arises. It takes some expertise to safely peck at a carcass on a major highway! Crows usually do it very well, knowing exactly when to take flight from dangerous traffic. However, young crows are vulnerable to accidents, as it takes time to learn to recognize the dangers of moving vehicles and at least a little good fortune to survive the learning experience. In any event, good fortune wasn't with Spirit.

As my brother Glenn drove east on Highway 103 near Halifax, he discovered this crow lying on the shoulder of the road, with his feet badly battered, his beak broken, and many of his feathers either broken or mangled. He was simply miserable! But he was fortunate to have been found by very caring people. In the months that followed, my brother and his family carefully nursed Spirit, then named Black Jack, back to a semblance of good health. Of course, little could be done for his damaged feathers, his broken beak, and his mangled feet.

This, then, was Spirit's condition when my brother gave him to me in October 1989. I was anxious to take him because I have always felt a deep affinity towards crows and ravens. As well, Glenn felt that the crow would have a much better life in a country setting, a place in some ways much less restrictive to a bird than the suburban environment in which Glenn and his family live.

Spirit was small for a young crow as his growth had been stunted by the accident. After giving the matter of creating Spirit's own space much thought, I decided to keep him in a large cardboard box in my cabin and was delighted when he accepted his new home. He particularly enjoyed the dried grasses in the bottom of the box. This was where he spent the remainder of autumn and the winter months that followed. In early spring, nature rewarded my two-winged friend with new feathers and mended his beak to its original shape and size. The latter was a miracle, occurring contrary to the educated opinion of certain people, who had assured me that a crow could not grow a new beak. His feet were another story. They healed but remained badly disfigured, making it impossible for him to perch on a branch. It was obvious that this crow could not survive on his own.

II

For weeks I felt the urge to name my crow companion. But I kept thinking, "Of what use is a name to a bird?" True, my brother's family had already named him, but I wanted something different, something which expressed a deeper meaning. For the crow's part, he certainly didn't appear interested in having a name. Humans are fond of classifying and naming things partly because we have this penchant for projecting our humanity beyond ourselves; we like to make nonhuman creatures conform to our ideas about them by assigning human characteristics or attributes. With this reasoning, we lose sight of the uniqueness of other species, such as cats, dogs, or crows, and it becomes much more difficult to appreciate and learn from species whose consciousnesses are relatively unknown to us.

Hence my reluctance to impose a name on my crow friend during the initial months of caring for him. I struggled with my desire to maintain the "wild" nature of the bird, to have as little influence on his development as possible. My attitude soon changed, however, as medical opinion confirmed that it was impossible to fix his legs and feet. I was forced to concede that this crow would live the rest of his life in close interaction with humans, especially myself, and I figured we were undoubtedly placed together for reasons unknown to both of us.

It was a short time later that a suitable name for him came to me. As I witnessed his effort to adapt to and overcome handicaps, to live a crow life within the confines of a human environment, I soon felt a tremendous respect for his spiritedness. For example, he quickly adapted to life within my cabin and to moving about in his straw box, contentedly pecking away at its walls. He was not averse to flying, despite more than one crash-landing. Eventually, he learned to land smoothly, despite his battered feet. Upon witnessing his determination, I realized that I should name him "Spirit." I thought this name was especially appropriate since, at some primal level, I felt he brought me closer to the spirits or devas, which legends say protect and nourish the minerals, plants, and animals on their evolutionary paths.

This helped me to realize what a blessing he was in my life. I believe I was touched by deva magic one chilly

November night, while asleep in the loft of my cabin. As I recall, I awoke to sounds of the movement of Spirit's feet on his grass bed. When I peered downstairs to investigate further, I was struck by sparks of moonlight dancing in his magical eyes.

It was as if the light of the moon was the medium by which something very special was communicated between us. I am convinced that this special "something" was the primal soul nature of the crow. If one accepts the premise that we humans are not alone in possessing the quality of soul, then it is not difficult to accept this interpretation. The world is, indeed, a magical place.

3

LIVING WITH A CROW

It was a wonderful privilege to share five years of my life with a crow, particularly one with the character of Spirit. It was especially interesting during the winter, which, on the South Shore of Nova Scotia, normally arrives by late November or during the dark days of December. Looking back, some of my most intimate moments with Spirit occurred during those long winter days and nights, when we shared the confines of my cabin, which is certainly small by most standards. In fact, the ground floor is almost exactly 14 ft. by 14 ft., with a loft above and an adjoining porch measuring 8 ft. by 10 ft.

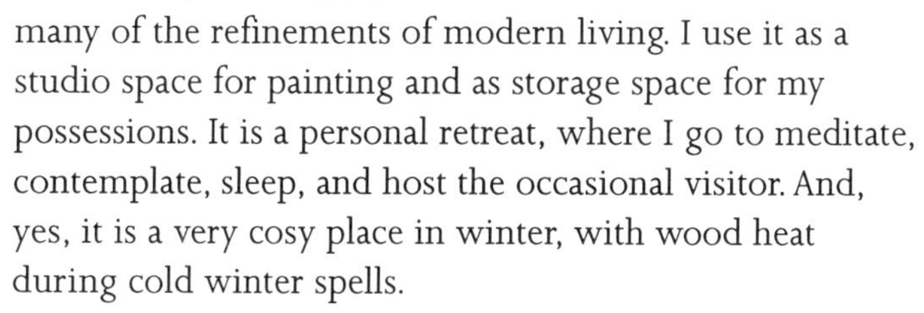

The place is basic, even rustic, and without many of the refinements of modern living. I use it as a studio space for painting and as storage space for my possessions. It is a personal retreat, where I go to meditate, contemplate, sleep, and host the occasional visitor. And, yes, it is a very cosy place in winter, with wood heat during cold winter spells.

During our first winter together, Spirit lived in a cardboard box, which I had furnished with a small bowl for water, a deep layer of straw, old spruce sticks, pine cones, and a variety of grasses carefully collected in the days of late autumn. The sides of his box were approximately a foot deep; in the front of the box, I cut a window through which he could poke his head whenever he was curious and wished to examine my cabin in detail. I quickly learned that a crow can deposit a mess of droppings in a few short weeks when he is confined to a limited area. I had to change his straw and grass roughly every three weeks to ensure his environment remained healthy.

However, the problem I most anticipated failed to materialize. I expected Spirit to routinely jump from his box and fly about the room in a wild, unpredictable fashion. He was certainly capable of such a display, but to my surprise, he preferred the comfort of his straw home, spending his energy contentedly pecking at the cardboard walls. As a matter of fact, on many occasions, I was awakened from sleep in the wee hours of the morning by his bursts of tapping against the cardboard. It became a familiar and even comforting sound and appeared to be Spirit's reaction to movements I made while sleeping (I suppose my snoring as well); it was his way of telling me things were okay downstairs.

Even as a young bird, Spirit was able to peck quite forcefully, which was why his pecking could waken me. I discovered that crows have very strong necks. This strength, coupled with much dedication and practice, enabled him to pepper his box with holes; by the time early spring arrived, the box was a mess. It looked as if it had been hit by a shotgun blast at close range! The holes were various sizes, many of which he could use as windows to poke his head through and look around. He had little use for the window I made in the box a few months earlier.

During the latter stages of this first winter, Spirit began to hide things—a favourite pastime of crows. I noticed the strips of cardboard, which he diligently pecked from the sides of his box, tended to disappear on a daily basis. Upon closer investigation, I discovered he was stuffing them under the straw and thick grass near the sides of the box. It seemed to be his favourite area for hiding things as he would regularly conceal food in those places as well. The food wasn't always permanently stashed away; sometimes he would collect it later.

Several people have written amusing letters to me regarding the crow's affinity for hiding and burying objects. For example, Norman Deale of Sheet Harbour, Nova Scotia, commented:

When Jim [a pet crow] was given more food than he could eat, he would always bury the remainder in the grass, or snow, for future use. Every spring he was busy reburying bits of food exposed by the melting snow.

Anything shiny attracted Jim, and that, too, would be carried off and buried—teaspoons, rings, etc. He was about one day when we were putting up an aerial for the old battery radio (a five tube Serenader). One of the white porcelain insulators, about 2 in. long and 3⁄4 in. wide with holes in either end, was lying on the ground. Jim apparently took a fancy to it and tried to carry it away for concealment. But every time he tried to pick it up, it would slip out of his beak. After several unsuccessful tries, Jim stepped back several paces, cocked his head sideways to study this unruly prize, then went up to it again. This time he arched his neck, stuck his 'top bill' through one of the holes, then closed his 'lower bill' on the end of the insulator! We watched as he tipped his head back then successfully ran off to bury another treasure!

Joan Stiles of Northport, Nova Scotia, wrote an account of a pet crow that loved to steal objects from people:

Jake [the crow] would steal anything in sight and hide it. I would be sitting outside on my lawn chair, having a cigarette and coffee. He would land on my chair and have my lighter gone so fast, then he would land on the telephone pole, put it down, and yell, 'Hello, hello.' What a time we would have to get it back.

I had to be careful about leaving a cigarette in an ash tray on the picnic table because he would steal it. We were afraid of him starting a fire. In fact, he almost did one day. I was outside on my lawn chair having my coffee and cigarette, when just as I was about to have a puff, that rascal snapped the cigarette out from between my fingers and went to the top of a tree, cigarette dangling from the side of his mouth, head cocked, looking smug. All kinds of coaxing could not make him give up the cigarette, until he made the mistake of yelling 'hello.' Down came the cigarette into the brush. If my husband hadn't been there, it could have started a brush fire. I am very careful since that incident.

Another stunt Jake pulled involved our next door neighbour, who came out to get in his car one morning, car keys in hand, when Jake swooped down, took his keys, and landed on our roof. The door bell rang and a very angry neighbour was demanding his keys. Needless to say, it took a lot of coaxing and many favourite treats to get Jake down with the keys.

We had calls from neighbours who claimed he was stealing milk money out of the milk bottles. He was accused of going in their children's bedroom windows and ripping their pillow cases. (This we didn't quite believe.) It reached the point that I hated to answer the phone because I might hear, 'Are you the crow lady?' I would have to wonder what our little black devil had done now? Maybe we made a big mistake when we spent hours throwing him in the air to teach him to fly, but we made a bigger mistake when we wanted him to be free. We wouldn't have done it any other way, though.

Stephen Czapalay of Barrington Passage, Nova Scotia, related some rather amazing incidents with crows. He wrote:

I taught mathematics in Quebec for thirteen years. On one warm spring day, I was teaching with the windows wide open. Suddenly, a crow appeared on a window ledge. I quietly whispered to the students not to move or say anything. Cautiously the crow entered the classroom, and from the window ledge it surveyed the nearby desks covered with math tools, pens, and such. It then hopped to the nearest desk, seized an eraser, and flew directly out the open window. The grade nine students there really enjoyed it, and after the crow's departure, the room buzzed with chatter. The crow returned several times to my lessons and stole a pen or a pencil each time. Then, sadly, we never saw it again. Imagine locating the nest of this crow—what treasures would be there!

In the same letter, Stephen recounted another amusing story about a mischievous crow:

About thirty years ago my brother and I were golfing at the Parrsboro nine-hole course in Cumberland County, a beautiful location beside the Bay of Fundy. He and I are both terrible golfers, and if either of us had made a half-decent shot, there would be a lot of talk and good-natured ribbing that followed. We had muddled through the first two holes—there's lots of forest and a high cliff on the course, so scores can run high if you slice or hook. Finally, there in front of me was the number two hole, about 200 yds. I teed off first. Amazingly, the ball landed on the green and rolled to within 2 ft. of the pin. Suddenly, out of nowhere, a crow landed on the green, cawed a couple of times, took several steps towards my ball, seized it in its beak, and flew off! According to my brother, I had to play another ball, which I proceeded to hook into the trees, to his great pleasure. Imagine finding the nest of this 'sporting bird' or a bewildered female crow sitting on the ball, wondering why it won't hatch."

II

By the time Spirit's second winter with me arrived in December 1990, he had grown too large for his cardboard box home. He still enjoyed the box as a retreat, but I had arranged his living space differently. I partitioned a space in my cabin as his private quarters, approximately 4 ft. by 6 ft. in area. The floor of this space was covered partly with masonite, over which I spread a thick layer of sawdust. This is where I placed his water pan, and the sawdust effectively absorbed much of the water, which Spirit frequently splashed from his dish. The remainder of his floor space was thickly covered with straw, and in a corner was the overturned cardboard box to which he could retire when necessary.

This was his favourite retreat if visitors entered the cabin. On those occasions, he swiftly hopped or flew to the box. Once inside, he would hide away and wait for the visitors to leave, or peck vigorously at the walls of the box, or perhaps peek out the door. Of course, his reaction to visitors often depended on his mood at the moment and the person visiting. For example, he was more apt to relax with members of my family whom he had seen before than with total strangers. Although the colour of their clothing also determined the nature of his reaction. For instance, Spirit would often react strongly if people approached wearing brightly coloured clothing, even if those people were otherwise familiar to him. In fact, he would react strongly even to me if I approached him wearing bright colours.

What really "fluffed" his feathers or annoyed him the most was when I came home at night and suddenly flicked on the lights in my cabin. The feathers on the back of his neck stood on end, while he gave me a severe verbal blasting with a series of low-pitched sounds, which I imagined were the equivalent of every "four-letter" word in the English language. He simply found it difficult to tolerate a sudden switch from darkness to light. I quickly learned my lesson and began to introduce light in a gradual fashion whenever I returned to my cabin in the dark. This practice proved to be much more satisfactory, and eventually he refrained altogether from verbal abuse.

III

"Crow as Art Critic" might have been the headline, had the local media caught wind of Spirit's penchant for fine arts. In case you think the following description is an exaggeration, I can assure you I sometimes laughed to the point of exhaustion over Spirit's antics while I painted.

He was totally preoccupied by my approach to oil painting. I recognized his fascination with painting quite by accident during our second winter together. One night I was painting a landscape from sketches that I had done earlier that day, when I noted Spirit's intense concentration on my canvas. It seemed as though his keen eye never missed a brush stroke. But what made me laugh so much was his response to my painting technique. If I became too rambunctious with the paint brush, he would commence cawing loudly. Often, I would experiment with moving the paint brush to discover which approach to painting was most pleasing to him. I found that if I painted in a slow, deliberate manner, he would be most apt to watch me without cawing loudly or becoming agitated. He did not react to bright colours of paint in the same way he reacted to brightly coloured clothing. This was probably because the paintings were small and unthreatening to him since they weren't animated objects, moving about in his environment.

Spirit also enjoyed country music! I wouldn't have realized this, except that I enjoy listening to good music while I paint. So, I noticed his interest in both painting and music at approximately the same time. Now, my musical taste varies depending on my mood. Sometimes I will paint to classical music, although there are occasions when I enjoy painting to the latest pop or country tunes. During that second winter together, I spent many late nights at the easel, listening to country songs and working into the wee hours of the morning. Spirit developed a great fondness for the hit, "Good Old Country Boys," recorded by Randy Travis and George Jones. He was especially pleased when I sang along, which was cause for great enthusiasm on his part. At such times, he would caw loudly and join me in singing the tune. In fact, whenever I wanted Spirit to caw, or make other crowlike sounds, I would sing along with Randy and George. Recognizing this fact was a stroke of genius on my part. I used Spirit's response to this song, and a few other country favourites, to elicit a wide range of crow sounds. It was easy, really. I simply sang at him, and often at close range. When I did this with vigour and emotion, Spirit would respond with a wide range of vocalizations. I think most bird trainers, birders, or bird-watchers would have been proud of my accomplishment.

Although the telephone is no art form, Spirit's response to the act of speaking over the telephone was even more extreme than his response to country music. Spirit's seeming attempt at interference made phone conversations quite difficult as he became vocal when he realized what I was doing. If I hadn't known better, I would have thought he was jealous of the phone! I don't know for certain what the motivation was for his response. Perhaps it was a social gesture on his part—he may have wanted to be part of whatever was happening. Whatever the case, his ruckus was often embarrassing; it was almost impossible to hear myself speak. Of course, the situation was less of a problem if there was an understanding person on the other end of the line, or if I was speaking to a friend who had previous exposure to Spirit's ranting behaviour. Not only the embarrassment but the humour increased if I was speaking to a stranger who was puzzled by a barrage of crowing songs in the background.

To solve the problem, I decided to take phone calls upstairs in the loft of my cabin, thinking the cawing would be less distracting. Well, it was but not to the extent I had hoped for because my cabin is small, and believe me, a crow's voice carries quite well inside a small wooden structure. Most worrisome were important business calls like the ones from my publisher. I especially didn't want Dorothy Blythe, the Managing Editor of Nimbus Publishing, to know. She phoned me at various times to discuss a manuscript I was writing then on plant medicines. At the time I didn't know her very well, and I wasn't sure how she would react to a crow joining our conversation. Besides, I was shy about explaining my crow to strangers because I was afraid people wouldn't understand the situation. Today, I would probably approach things differently and wouldn't be as paranoid about Spirit interrupting my phone calls.

On one occasion when Dorothy phoned, Spirit was feeling especially ornery and cantankerous. He began ranting and crowing to sweet heaven the moment I picked up the receiver and said hello. It was terrible! Huge beads of sweat began to break out on my forehead. Frantic, I dashed headlong for the pile of blankets on my bed and dove under them, phone in hand. Piling on top of me everything I could grab within reach, I tried to bring some semblance of normality to the situation. All the while, of course, I endeavoured to talk to Dorothy in a calm, level-headed manner. The heat was awful with the half dozen heavy blankets, pillows, shirts, and other assorted items over my head. Comical in retrospect, it was a very tense experience at the time, and one I wouldn't care to repeat.

However, as the years passed and I learned more about the ways of crows, I improved my sense of humour and learned to flow with the occasional novel situation. I adjusted to having Spirit in the cabin and eventually didn't give a hoot whether he crowed when the phone rang. In the meantime, I did discover a method whereby I could have quiet conversations if I wished. The purchase of an answering machine allowed me to collect my messages and answer important calls from a pay phone or another residence. Later, I made the important discovery that Spirit would rarely caw in the darkness. Hence, I was able to have quiet, peaceful phone conversations after sunset, with the lights off in my cabin.

IV

I heat my cabin with a wood-stove during the winter, and Spirit seemed somewhat fearful of its size and its doors. On many occasions, when I opened them, he would caw loudly and head for the safety of his box. It seemed the movement of the doors would alert him to the possibility of danger. However, this fear would pass quickly, and he would become quite excited when I prepared a fire, although I could never determine whether he preferred a well-heated room or a cool room. I didn't notice any great difference between the two in his activity level or vocalization pattern.

During our years together, Spirit rarely had to contend with winter cold. Once, he was caught outside in very cold temperatures during the latter part of November. The weather had been quite mild up to that point. The forecast called for continuing mild temperatures. I had no idea that it was going to change so drastically. Winter arrived suddenly that particular night. The temperature dipped very quickly, accompanied by snow and high winds. As Spirit hadn't been exposed to such temperatures before, I was very concerned. Rushing outside at two o'clock in the morning, I was amazed to find Spirit sitting calmly, facing the wind and snow with no visible sign of discomfort. I had to make a decision rather quickly either to leave him where he was or to bring him into my cabin. I chose to leave him outside, mainly because I knew how defensive and ornery he became during the hours of darkness and it would be difficult to catch him. Since he didn't seem the least bit perturbed by the weather, I decided it wasn't the emergency situation I had envisaged. The next day I carried Spirit into the warmth of my cabin. Florence Hubley of Tantallon, Nova Scotia, confirmed in a letter to me that crows hate to be disturbed at night. She wrote, "*We covered him over at night with a box. He would snap at us if we touched him.*"

I have always felt a sense of awe over the way crows and other wild birds are able to survive in extreme winter weather. Scientists claim that the black feathers of the crow and raven are a distinct advantage to them in cold temperatures, since they absorb short wave energy, decreasing the temperature gradient between the skin and outer feathers. This makes those birds well equipped for Canadian winters. Even so, knowing how I had pampered Spirit through the winters of his life, I worried about how he might weather his first storm. You can imagine my relief when I discovered how well he handled the temperatures of that particular November night. I was prepared to make an emergency rescue and could probably have managed it, despite some difficulty, since I had noticed earlier that if I shone a light in the area of his living quarters, he would quickly regain daylight consciousness, and lose his hostility towards me. I imagine Spirit's defensive attitude at night was genetic, an inheritance from generations of crows, who were vigilant against attacks from predators, such as owls, that search for food during darkness.

4

THE CARING

Recently, a friend remarked to me, "Oh, it must have been so romantic and exotic to have a crow as a pet!" Well, this is true I suppose, after all it's certainly unusual and does give one the opportunity to learn all kinds of things about crows. However, looking after such a large bird can entail much hard work, especially if you have limited resources and have to care for the bird over a long period. For one thing, it's very time consuming if you wish to give the bird a stimulating and rewarding life.

I tried to ensure Spirit had exercise, a caring environment, and things that stimulated his interest. I tried to remain open and alert to learning new lessons from him. When I think about Spirit's effect on my habits, I realize how deeply he influenced my life; I am amazed at the living adjustments I made to meet the numerous challenges that arose during our five years together: adjusting quickly to the threat of predators, creatively solving the problem of talking in peace on the telephone, or taking time on the spur of the moment to interact with Spirit when I sensed he was feeling bored or lethargic. Of course, we should be ready to make time for any pet, whether it is a cat, a dog, or a more exotic creature such as a crow.

Spirit had ways of increasing the work involved in his care. His water supply was often difficult to contain, especially when Spirit was indoors. For example, when he decided to bathe by splashing his beak about in the water, he would quickly wet a significant area of his living space. It wasn't so difficult in the warmer months, when he was outside. Then, a couple of basins of water adequately met his needs if they were changed frequently, and, believe me, there were instances where they required frequent changing. On occasion, he would mix earth and straw together in his basin, resulting in a rather thick, dirty pool of water. When he was on such a roll, he became an involved and excited bird, though somewhat demanding. He would approach the basin as soon as I poured the fresh water and begin a familiar ritual with his beak, moving it to and fro through the water so that most of his beak was immersed, and with every few strokes he would wipe it on the rim of the basin. When in the mood to "work" in the dirt, Spirit would often peck or stab at the ground, filling his beak with earth, which he then promptly dumped in the basin. He ritually wiped his beak prior to collecting the next batch of soil.

A crow's beak is a wonderful tool, which they use to good advantage. I received a note from F. Weatherbee describing a rather remarkable experience in which a crow made ingenious use of its beak.

While my neighbours complain about the 'squawky crows,' I encourage them with all sorts of goodies to eat. On one such occasion, I threw out a dozen stale crackers, and before they even hit the ground, a crow appeared and started strutting toward them. When he was convinced it was safe to take one, he picked it up in his beak and then laid it down. He moved on to pick up the next cracker and put it on top of the first one. Then he picked up the two crackers. As before, he laid them down and proceeded to the third one. Again, he went to the pile and stacked this on top. I was amazed to see him pick up a stack of four crackers and fly away with them. He had tried to add a fifth cracker but gave up when he realized it was too much to handle. He did return and he brought some friends, but they settled for a snatch and grab lunch, nothing as spectacular as their friend.

One day, Spirit was particularly intent on working the ground—in fact, fanatical better describes his behaviour! He splashed, dug dirt, and made his water dark and swamplike in a matter of minutes. I was feeling like a trickster myself so decided to change his water as fast as he dirtied it. What a time we had! He was very industrious. If I attempted to intervene with crow talk, he would stop momentarily, look my way, reply, and proceed with his work. This continued for quite a while as I changed the water six times. Suddenly turning about, Spirit hopped off in the direction of a mound of peanuts piled in a corner of his outdoor pen. He wasn't in the least perturbed that I had so frequently changed his water or that I saw humour in the situation—I was laughing up quite a storm. Rather, he simply ignored me as I sat in his pen, watching him with a silly grin on my face.

II

In their natural setting, crows seem to thrive on touch between one another. This sense of touch shouldn't be ignored if one wishes to build a caring and lasting relationship with a crow. I have often observed crows grooming their feathers or the feathers of their mates. Ravens, too, are like crows in this respect. In fact periodically, a pair of ravens frequent the maple tree near my cabin, where they groom each other regularly.

Spirit had a weakness for grooming; he was open to such affection or quality time on a daily basis. He especially enjoyed our "beak scratching" sessions, when I would scratch his beak for long periods of time, particularly the base of his beak, where it met the flesh and the growth of short, fuzzy feathers. He would usually lean forward, resting his beak on the ground to support his balance. The scratching and massage had a hypnotic effect on Spirit. He closed his eyes and seemed to enter a sleep-like state. At those times I would take his head in my palm and give it a total massage. I could continue the massage for as long as I wished—he loved it. He was such a sook!

I'm convinced that "play" is important in bird life, and that it is instinctive in bird behaviour. It certainly must be considered when creating a loving, caring environment for domestic or wild birds. Crows have a wonderful sense of play. I used this to good advantage in creating games or situations where I could interact with Spirit.

One game involved using his old broken feathers. Spirit was prone to broken feathers, partly the result of spending so much of his life on the ground. His tail feathers, especially, tended to break, so that they were in poor condition by the time he lost them in the spring. So, I invented this interactive game based on the resolve that whenever Spirit lost a feather, I would stick it in the bark of an old piece of dry wood, which I placed in his box. At one point in the early spring of 1991, I had a half dozen broken feathers in the piece of wood. They were placed in a row, in a decorative fashion, and looked quite attractive from my perspective.

However, the arrangement proved temporary. Arriving home one evening, I found it completely disassembled and wrecked! Spirit had tossed the feathers from his living quarters. Soon, thereafter, I repeated the arrangement with the same results. Spirit would grab the feathers with his beak, pull them from the wood, and fling them across the cabin floor. I continued the game for a couple of weeks on practically a daily basis, always arranging the feathers in a neat row, only to have Spirit pull them from the wood as often as I placed them there.

Crows seem to enjoy moving objects around—even playing with them—as my feather game with Spirit suggests. Joan Stiles comments on similar behaviour in Jake, a pet crow:

One of his most devilish deeds was swinging upside down on our neighbour's clean clothes hanging on the clothesline. She would come running out with her broom to chase him away, and he would just hover over her head and taunt her. When she went inside, he would take the clothes pins off the clothes, let them drop to the ground, and then play in them.

Even if I presented Spirit with a single feather, he would grab it from my fingers and toss it away. I was surprised at the strength in his neck, revealed in his grabbing and tossing. He was really quite funny! He made me laugh, a deep gut laugh—the kind which feels really good and brings tears to the eyes. We should all learn to cultivate this kind of laughter, it's good for the soul, and makes the world a more joyous place. It was like a gift that Spirit gave me, this laughter, and it was nice for me to know that he received plenty of excitement in return. It was at this point in my life that I realized the truth of Henry Ward Beecher's remark, "*If people could be feathered and provided with wings, very few would be clever enough to be crows!*"

5

CROW TALK

Crows are marvellous birds, and my time with Spirit taught me many things about their world, none of which is more outstanding or surprising than the repertoire of sounds which have specific meanings for conditions or events in their lives, and what I refer to as their language. The reader may smile over my use of the word "language" to describe crow vocalizations. But, if you're ever fortunate enough to become intimately familiar with crow sounds, you will likely feel more inclined to accept my broad definition of language.

It is difficult to express the excitement I felt whenever Spirit did something unusual or vocalized in a way I considered foreign to crows. For example, have you ever heard a crow chirp? Unusual isn't it. Yet, on occasion they do chirp, which is a sound you will probably never hear from crows in the wild, when observing them during casual walks in the forests or fields. You have to develop a close relationship with a crow before it will reveal the full extent of its vocabulary, and since crows are not greatly loved by the majority of people, few have heard the more esoteric sounds of crows. I was fortunate to have Spirit as a teacher.

Of course, before Spirit came on the scene, I was like most people, not paying much attention to the vocal behaviour of crows. All the cawing sounded similar to me. For example, I didn't recognize (nor was I aware of) the "alert" or "beware" call a crow normally makes when humans or other creatures are in proximity. This is a rapid series of rather short, high-pitched caws, which are easily identified with a little experience. Those rapid calls should not be confused with the call of a young crow, which may sound like "car, car, car." I first heard Spirit give the alert call when he spotted my brother Garry's cat crouching nearby in the tall grass. Spirit was furious, and in a minute or so the alert call changed to a flurry of angry caws directed at the startled cat, which soon turned and left the area. After that incident, Spirit and the cat got along just fine.

If you listen closely to cawing, you will notice how physical circumstances produce slight variations. For example, if you walk past a crow sitting on a tall pine tree, you may elicit a different response from what you would receive if you approached the same crow with a dozen of your friends! Often the crow will simply fly away; however, if you want to be a serious student of crow language, you have to pay careful attention to both the frequency and the pattern of cawing, and to the pitch or tone, and to the urgency or intensity of the sound. I am now somewhat sensitive to those things and have noticed there are occasions when crows caw loudly, simply for the sake of singing to the world!

Francis Henry Allen, a man who was very keen on crows and appreciative of their language and behaviour patterns, wrote with great persuasion about their aesthetic sensitivity in an essay titled "The Aesthetic Sense in Birds as

Illustrated by the Crow." He noted the wonderful "time rhythm" of their cawing and observed that while the caws are frequently in triplets with a rhythm of 2-1, there may be four caws in groups of two (2-2) or in the rhythm of 2-1-1, very regular and without the slightest variation. While F. H. Allen did not suggest those caws represented language or signals, he didn't dismiss them as purely mechanical either. He believed that crows take much delight in their ability to utter a variety of rhythmic songs or sound patterns and that in a limited way, the crow is a true artist, composer, and performer. Certainly, Spirit was very expressive in my presence, uttering notes that ranged from annoyance, to anger, to more pleasant melodies indicative of humour, joy, and play. J. P. Porter noted that a crow can learn to open a door simply by watching another bird that had been trained to do it. The crow's ability to remember allowed it to apply what it had earlier observed.

Spirit was also very good at mimicking, mocking, and replying to sounds. Over the years I learned to communicate with him either vocally or through "tapping" sequences. For instance, in the winter when Spirit was in my cabin, I would often listen for his movements when I awoke during the night. If there was silence, I would tap my knuckles three or more times on the floor, then listen for a reply. Spirit was somewhat of a trickster and might remain quiet, though I knew he was listening. So, I would tap again. Soon he would respond by tapping his cardboard box or some other object, sometimes with an exact copy of my pattern. When this happened, we usually continued our game until we became bored, or I fell asleep.

Crows in general are excellent mimickers. I received letters from several people who attest to this skill in crows. Norman Deale remarked:

> *Before he was a year old, Jim [the crow] had learned to imitate me. He would sit on my wrist while I repeated, 'Hello,' over and over again. Soon Jim was replying and using my own tone of voice. Besides, he laughed exactly like I did—an amused sort of giggle, as I recall. And he also learned to call, 'Help!'*
>
> *The surprising thing was that Jim seemed to have reasoning ability. When the dog or young kids threatened him, you would hear him calling, 'Help!' He would sit in the trees beside our lane and say hello to anyone coming to the house. Many people would return his greeting, and then he would laugh, much to their embarrassment.*

Myrna Wilson of Upper Rawdin, Nova Scotia, wrote about a crow that learned to say his care-giver's name:

> *Hoppie [the crow] could talk. He would say a number of things and would sit on top of the house and chatter away. Sometimes it was easy to make out what he said. However, the word he used the most was 'Elmer,' the name of the boy who looked after him. Hoppie would follow him to school and sit on top of the school and holler, 'Elmer' as loudly as he could.*
>
> *The teacher contacted his mother, asking that they keep Hoppie home. They locked him in the porch while Elmer walked to school. Later, they released him, and it wasn't long before he arrived at school again, hollering, 'Elmer.' Everyone thought he was so smart to find his way to school all alone.*

Florence Langille of Tantallon, Halifax County, wrote about a crow that learned to mimic chickens clucking:

My brother Donnie, who now lies in France, after losing his life in the very early days of World War II, was an avid woodsman. Each spring he brought home a young crow. They would spend the summer with us, returning to their own gang in the late fall. We never confined them and always let them leave when they were ready. Only once do I remember one spending the winter in the henhouse, having, I guess, decided he was one of them.

Each crow was for some reason called Joe. They would all learn a few words by the time summer was over, mainly 'Hello' or 'Hello Joe,' which were the words they usually heard. But they picked up other sounds. I remember my mother searching in the hedge for a clucking hen, figuring she had hidden her nest away to hatch some chicks, as they liked to do. After pushing the bushes aside, who walked out but that summer's Joe, happily clucking as he had heard the hens do.

Finally, Madeline Way of Dartmouth, Nova Scotia, commented on her neighbour's (Raymond Swinimar's) crow experiences:

A few years ago Ray had a tame crow that learned to say, 'Hello,' very clearly and precisely. As he grew older he flew away with other crows but remained in the area. You can often hear him in the trees saying hello, but not as clearly as he gets older—he never caws as other crows do.

Right now Ray has another crow who, as soon as he hears Ray's car, comes flying down to land on the car window. As Ray emerges from the car, the crow lands on his head. Jo Jo [the crow] has many little tricks, such as clinging to the wiper blades and going for a ride.

II

Crows are excellent "wake-up birds," my own term for birds of all kinds that sing exuberantly early in the morning. Spirit ranked among these; he would commence a loud cawing routine if he heard me stirring or moving any time after daybreak. I was aroused from beautiful dreams on more than one occasion during our five years together. If I was able to lie very still and if my luck held, Spirit would give me much needed peace and quiet during the early morning hours.

Earlier, I mentioned the chirping sound, which Spirit made on occasion. I was very surprised when I first heard him make the sound. It was during a winter evening, and I had just returned from a trip into Bridgewater. When I entered the cabin, I was greeted with a low chirping voice. Eventually, I learned that it is a "greeting" in crow language. Barely audible, it is easily missed if you're not listening carefully. I wasn't even aware of this sound during the first three years of Spirit's time with me. Perhaps it is an expression which crows develop in later life. Or, perhaps I simply didn't notice it earlier, or he simply didn't use it to greet me in that intimate fashion. Spirit also made a low "peeping" sound, which, like the chirp, appeared to be a greeting gesture. I used to hear him make this sound at night. It always occurred when I moved about in the darkness of my cabin. For example, if I came downstairs late at night without using lights, he would often greet me in this way. It was clearly a greeting sound or recognition of my presence.

On occasion Spirit made a low cackling or cracking sound, accompanied by a high-pitched vocalization somewhat akin to "wa-ha-ha-ha." Those sounds are made with the wings slightly spread and the tail feathers vibrating or moving rapidly. I noticed that if I imitated this sound, he would move his tail feathers and respond vocally. Apparently, what I call a cracking sound is quite similar to the courtship song of the crow, which is described as a rattling sound, or, "a quick succession of sharp notes which have been likened to the grinding of teeth," according to Arthur Bent's *Life Histories* (p. 226 ff.).

Perhaps Spirit was giving me his best courting performance! The rapid movement of his tail feathers and the spreading of the wings were certainly significant behaviours. Whatever the motivation, it is safe to say that when a crow makes such a performance, it is expressing a willingness to be approached and a need for affection. As noted, Spirit enjoyed receiving affection, especially having his head and beak scratched. He would frequently spread his wings when I massaged his head.

In fact, you may wonder what position I held in Spirit's eyes. Well, speaking in terms of the crow's perspective, I speculate that my position shifted between crow parent and partner. In some respects, perhaps Spirit never grew up, and this was reinforced and reflected in his dependency on me for his nourishment. He certainly responded like a young crow whenever I fed him. He would look up at me and open his beak as a young bird would. I would either put the food to his beak or, more often, simply place it on the ground or in a dish near his water container. So, in terms of food, I seemed to play a parent role. In terms of his behaviour, whenever I spent time with him in and around his living quarters, I seemed to fulfil the role of mate or partner because he would respond to me with that rattling or cracking sound, accompanied by spreading his wings slightly and moving his tail feathers.

I would highly recommend the study and observation of crow talk to anyone who has an interest in nature, bird life, or wildlife in general. It will provide years of discovery and enjoyment. The thrill of hearing a variation in crow vocals for the first time and of gradually learning the meaning of various sounds is worth the dedication required. You may better appreciate the process of communication among creatures other than humans.

CROWS AWAY: Observations on Crow Flight

I have observed the flying behaviour of crows for several years but especially when Spirit came into my life. If you are patient and dedicated in your observations, over a period of time you will learn much about their behaviour and flight patterns. Of course, it is important to observe the crow in many environments and at different hours during the day. Early in your observations you will appreciate the truth of the old adage, "As straight as the crow flies," which, as I learned, is a regular flight pattern for crows traversing fields or large bodies of water. Accompanied by steady, monotonous wing movements, this flight style could be considered boring by some, but when I see a crow flying in this manner I am inspired by its strength, endurance, and sense of purpose.

The blue jay is frequently seen in the company of crows, whether in flight or on the ground. Somehow, I doubt whether crows enjoy the companionship of this relative, as the jay is a pesky bird and occasionally takes delight in diving at the more cumbersome crows. Seeming to enjoy this exercise, jays will gang up on a single crow and make life quite miserable for the poor bird. The blue jay is a very intelligent bird in its own right and may seek the company of crows, which are adept at finding food, the jays are pleased to lend a beak in devouring it.

I'm convinced that the crow doesn't receive enough credit for its flying ability. We continually hear about the wonderful flying ability of the eagle, osprey, or hawk, or the swiftness of the falcon and the sparrow. Of course, with respect to flight, those birds are in a class of their own. Nevertheless, the crow deserves its due. Spirit demonstrated a good deal of speed, strength in flight, and an excellent ability to manoeuvre among trees at low flight. Also, the acrobatic manoeuvres of two or more crows when they play in the sky is a beautiful sight. Millicent Flicken wrote an essay titled "Avian Play," in which she describes witnessing a crow trying "a variety of half-turns, walking in air, and partial slips and rolls," and was convinced that the crow was playing, even demonstrating a delight with life. Another writer, Fred Pierce, describes in "A Crow that Nearly Looped the Loop," a humorous and unusual incident involving a crow. Apparently, this crow was flying overhead carrying food with his feet, when he attempted to transfer the food to his beak. In the attempt, he bent his head so low that, losing his balance, he almost did a somersault in flight. Lucky fellow—not the crow but Fred Pierce, who was fortunate enough to witness the event.

Recently, I observed another characteristic of crow flight while walking the eastern shoreline of Minamkeak Lake. A crow was flying north of me, about 100 yds. from where I was walking. It was near the tree line when it suddenly veered and began moving southerly into a strong headwind. As it flew over the waters of Minamkeak, I noticed its open beak, as if it were gasping for breath. Having seen Spirit display similar behaviour, I realized the crow wasn't exhausted but was simply reacting normally to a situation requiring great exertion. I watched that crow for several minutes until it was invisible to the naked eye. With my binoculars, I could still detect its open beak and its rhythmic wing movement as it approached the hardwood hill on the far shore.

II

Surprisingly, Spirit's flying ability was not greatly affected by the injuries he suffered as a young crow. He could fly as fast as most crows, although his stamina probably was not as good as that of crows in the wild. On occasion he had difficulty landing because of his lame feet. At such times he would land, take a tumble, and usually come to rest upright on his feet. It was quite funny to watch his unexpected agility that compensated for a lack of gracefulness on the ground. Generally, he became quite adept at landing, as long as it was on the ground rather than in a tree.

I will always remember 15 May 1991, because on that day Spirit had a rather remarkable flight. He had flown often in the past, but never with such eventful results. It was a beautiful, warm spring day, and there was a steady wind blowing from the southwest—a mild, inviting wind, which Spirit would quickly learn to use to good advantage. If I had foreseen the eagerness with which he would fly, I would have hesitated giving the gentle push that released him into the breeze. In fact, it was hardly a push; I simply held him with my hands, moved them forward and up, and he was away, flying strong. I was surprised at the strength of his wings, the certainty of his flight, and the speed with which he moved from me.

He flew low at first, gradually gaining altitude as he crossed the field. I had expected Spirit to cross the field, glide low amongst the tall hemlock and pine trees before coming to rest as usual on a floor of familiar pine needles. However, this flight was different. Upon reaching the tree line he rose sharply, a good 80 ft. above the ground, so I knew he wasn't going to sit on those pine needles. He soared as the wind lifted his body up and over the tallest hemlock, and in a few moments he was gone. It is difficult to describe the emotion, the feeling in my stomach as he disappeared from sight. I felt joy! I was sad. I felt ecstatic! I was lonely. I thought, "There, you idiot, now you've done it. That crow's on a one-way ticket out of here!"

I tried rationalizing the situation to convince myself the best thing had happened; it was appropriate for Spirit to fly away and live the normal life of a crow, surviving as long as possible in a natural environment among those of his own kind. Yet, despite my positive thoughts and rational approach, the deep empty feeling would not go away.

This internal dialogue was short-lived, as I quickly collected my wits after remembering that Spirit was at greater risk than other crows because of having to land on the ground. I immediately began my rescue efforts. I had visions of Spirit sitting in a wooded area, falling prey to a raccoon, a cat, a fox, or some other predator, which could easily sneak up on him on the ground and devour him before he even became aware of its presence or could get airborne. At the time, I believed those terrible thoughts were realistic because, after all, if a bird is pursued in a heavily treed area, it is difficult to elude predators, especially if that bird has trouble perching.

I must have been a comical sight charging over the field in pursuit of that crow! I stormed the clearing, carrying a cardboard box under my left arm, and franticly shouting, "Spirit!" as often as my breath would allow. Spirit was accustomed to riding in such contraptions: it had become my practice to bring him home in this way after our excursions in the woods. Left to his own discretion, he would remain there for hours. My schedule just didn't allow me to remain with him for such long periods of time and I certainly couldn't leave him there on his own, since his injuries prevented him from perching in the safety of trees. I felt that a crow on the ground was fair game for many kinds of predators. Therefore, I simply had to carry him home and discovered that a cardboard box was best for this purpose, since he disliked being carried in my hands. (He didn't like having hands wrapped around his body.) The box was especially good if I had to carry him any great distance.

Entering the tree line, I paused, took a few steps, then paused again, looking and listening for crow sounds or other clues to his whereabouts. There was a stillness, a silence, except for the chirping of small birds on birch trees nearby. Several large, soft looking clouds floated in the sky, and my eyes were attracted to streaks of sunlight scattered through the dark shade of conifer trees. I found myself imitating crow calls as I walked deeper into the forest, all the while gazing at the ground, or upward at the tall trees as I madly searched for Spirit. I must have walked a quarter of a mile in this fashion, before pausing again to consider my strategy.

Anyone who might have witnessed the "crowing" sounds I made that day, would have considered me borderline crazy! I still wonder if anyone heard me. I ßsuspect that if there was another person in that forest, they would have made fast tracks out of there, rather than attempt to discover the source of such a racket. I know that if I had been in the other person's shoes, I certainly would have done so!

III

As I stood still, rethinking my so far unsuccessful approach, I distinctly heard the sound of a crow behind me. It was faint but familiar, carrying the echo of Spirit. Over the years I gained some familiarity with Spirit's vocals and could usually distinguish them from other crows' sounds. In this instance, because the sound was faint, I couldn't be sure it was him, especially since I knew there were other crows in the vicinity that frequently visited this forest.

Quickly retracing my steps towards the field, I listened attentively to the responses to my calls, which convinced me the crow was Spirit—it certainly sounded like him. I soon spotted him, sitting beneath a hemlock, furiously poking the ground with his beak. To my surprise, he was near the edge of the field; he must have circled back to the comfort of his home territory. I had obviously underestimated Spirit's flying ability and navigation skills.

My lack of confidence stemmed from knowing he had little flying experience compared to crows in general, having lived much of his life in a confined space. I was convinced that Spirit would easily become lost during free flight, having failed to consider the genetic inheritance and keen sensory capabilities of a crow. Remaining in or relocating familiar surroundings was probably an easy accomplishment for Spirit.

When I attempted to pick him up and place him in the box, he quickly hopped away from me. It was as though he had read my mind and definitely wasn't having anything to do with my plans. So, I sat under a hemlock with my back to the trunk and watched Spirit dig away at the ground. After twenty minutes or so, I was able to place him in the box and carry him to the more familiar surroundings of his living area.

Several things remain unclear—was he under that hemlock when I began my search? Was he watching me as I made the mad dash over the field, wildly calling his name? Perhaps the temptation to lead me into the forest was too strong in him to ignore. The trickster in Spirit wanted some fun! I suppose this can be attributed to the legendary character of crows, after all, they have been performing such antics since the mythic past.

7

A CROW'S DIET

Crows are renowned for pecking through garbage and eating all kinds of things tossed aside by humans. Thanks to Spirit, I had the opportunity to examine more fully a crow's dietary choices. In this respect, I was careful to compile notes on Spirit's food preferences and general eating habits. Of course, he had ample opportunity to sample foods, which the ordinary crow may only dream about. He was able to ignore and show his distaste for certain foods, as there were always several options; in other words, he was spoiled! So, Spirit's food choices cannot be fairly applied to a crow's choices in the natural environment; however, I feel confident that, given the opportunity, most crows would have likes and dislikes similar to Spirit's.

For instance, Spirit adored cucumbers! I would frequently toss a chunk of cucumber to him, and he would eagerly proceed to hollow it out, leaving only the shell. He was especially fond of the seeds, which were the icing on the cucumber cake for him, a delectable item to be savoured, slurped, or swallowed whole. The cucumber was Spirit's favourite vegetable, if one is to judge food by the amount of enjoyment taken in devouring it.

In the summer, when the blueberries ripened, I would usually take Spirit to the field, where he could eat freely among the berry bushes. He was adept at picking blueberries from the stalks and capable of cleaning a bush of its berries in a matter of minutes. Feeling mischievous, I once placed a full quart box of berries in his pen. The result was hilarious. There were berries everywhere—Spirit upset the box and was literally bathed in blueberries. I was afraid he would overdose! Fortunately, he was able to eat his way around the problem, so to speak.

Grapes, too, were a favourite fruit of his. There is a huge grape stalk near my cabin, a very old variety, producing medium-size dark purple grapes. They are somewhat tangy but have an excellent flavour and make wonderful wine and jelly. In the autumn, when the grapes turned a deep purple, I would place Spirit in an open, low-sided box and set it among the vines in easy reach of the fruit. I would leave him there for awhile, so he could enjoy the natural environment and the all-you-can-eat buffet.

He ate a variety of nuts, although peanuts were his favourite. He had two methods for cracking the peanut shells. Often, he would stab the shells with the point of his beak. The other method was to pick up the peanut shell and crush it in his beak. He was very clever with his beak and shelled peanuts quite easily. Occasionally, I would shell them for him if I thought he needed a quick treat.

Of course, Spirit was fond of almost any kind of meat but he had an especially good appetite for hamburger, bologna, and roasted chicken. Raw hamburger was probably his favourite, although he relished pecking the meat from the bones of baked or fried chicken. As for bologna, he preferred Larson's waxed variety, as do I. Oh, don't get me wrong, I'm not a big eater of bologna, but there are times when a mustard and bologna sandwich tastes good.

I've always known that crows had a soft spot for the taste of fish. When Spirit came upon the scene, I quickly discovered that he loved fish oil capsules. It didn't matter whether the capsules contained cod or halibut liver oil—they were both favourites. The capsules are bright, which is reason enough to attract a crow. Add to that a fish taste, and you have crow heaven. Spirit would play with an amber-coloured capsule, moving it around in his beak and squeezing the capsule until it broke, squirting oil over his beak and surrounding area. The oil was beneficial to his skin, and if he took it regularly, I noticed his feathers had a particularly beautiful sheen.

If there's anything a crow likes almost as well as fish oil capsules, it would have to be cheese. In fact, both crows and foxes are legendary lovers of cheese. Witness the folktale about the fox and the crow and how the clever, devious fox tricked the crow into dropping the cheese. In this folktale, the fox compliments the crow on what a wonderful voice he has and on how beautifully he can sing. Becoming very proud of himself and wishing to show off for the fox, the crow starts to sing a song. In the process, he drops the cheese to the ground, where the sly fox is all too willing to relieve him of the delicious treat. Spirit loved cheese, and I think any clever old fox would have had a rough time getting it from him. Regardless of the form—hard cheese, cheese sticks, cheese-flavoured crackers, or macaroni and cheese—Spirit loved it; cheese was a delicacy in his diet.

II

A crow enjoys water—lots of water—to splash around in and to wash its food. It is as essential to the well-balanced life of a crow, as it is to the balance and harmony of most creatures who enjoy playing and bathing with water in eating rituals. I must admit that I was somewhat taken aback when I first discovered that Spirit liked to wash his food. I accidentally placed his food, hamburger, very close to his basin of water. I watched him collect the meat in his beak and wash it in the water. I was reminded of the raccoons, which are famous for this habit and will wash their food at every opportunity, but I didn't expect a crow to do the same thing. Yet, here was Spirit taking chunks of hamburger, placing it in the water, then immediately collecting the meat in his beak, and shaking it a couple of times before settling down to a meal. I later learned that he would perform a similar ritual with a variety of foods, including bread, bologna, and cheese.

Several people have commented to me about crows washing or soaking their food in water. Iola Stronach of Kingston, Nova Scotia, wrote:

It seems this crow can't caw. It makes a very strange noise. I've remarked to my husband and others that it has a sore throat. One day it came to the birdbath with part of a cooked lobster shell. It dipped the shell in the water many times, turning it over and over. When he or she was satisfied, it proceeded to remove what small amounts of meat it could find in the shell. Then it removed the meat from the water, while my daughter and I watched.

The next day I saw it land on the birdbath with a slice of bread. The bread looked stale and hard. I couldn't believe my eyes when it placed the bread in the water, turned it over a couple of times, removed it to the grass, and ate it. I always thought birds were stupid. Now I think they are darn smart! I wonder if it really did have a sore throat.

Doris Phillips of Halifax, Nova Scotia, commented on similar behaviour in crows:

My husband and I saw a crow come to our birdbath with a large white object in his beak. He placed it in the bath and immediately flew away. I went to see what it was he had left behind and soon discovered a good-size part of a loaf of hard, crusty French bread, which was soaking in the water. In a short time, the crow returned and commenced eating the softened bread. Good thinking, don't you agree?

Other crows have been partial to either making use of water in the process of eating or showing a preference for wet or moist foods. Norman Deale remarked that Jim the crow "*thrived on a diet of home-baked bread dipped in warm milk and supplemented with earthworms dug from the garden.*" Florence Hubley of Tantallon, Nova Scotia, wrote that her crow Jody's "*favourite meal was raisin bread. He would pick the raisins out and leave the bread. He also liked a dish of porridge in the morning.*" The porridge is consistent with a crow's apparent love of wet foods. Finally, Sheila Wilson of Pugwash, Nova Scotia, stated that Joe the crow ate from a small spoon "*pig feed or oatmeal mixed with warm water, which he devoured with a gobbling, squawking sound with every bite.*" As he got older, like Norman Deale's crow, Joe supplemented his diet with worms, which he reportedly ate with "gusto."

I discovered from Spirit that water was not just for food but for fun, as well. I normally moved Spirit to his outdoor living quarters during the first week of April. In the spring of 1992, after placing him outside, we were hit with a late season storm, which dropped a 4 in. covering of light, fluffy, "robin" snow. I made snowballs and gave them to Spirit, which he enjoyed thoroughly. He would peck at the snowballs and move them around with his beak. On one occasion, he placed a snowball in his basin, moving it through the water until it broke into several pieces. Spirit played with that snowball until it was only a thin covering of slush over the surface of water in his basin.

Spirit liked to play with bones in the water, too, especially small chicken bones. He appeared to be fascinated with the way they moved through the water. He would push the bones with his beak, eventually plucking them from the water, and pecking them clean of meat. He was particularly fond of washing grapes in water. He would drop a grape in the basin or hold it in his beak and move it back and forth through the water. On occasion he attempted to swallow the grape whole; but, more often, he squeezed it until it burst and savoured the juicy bits one by one.

Spirit's eclectic diet was no great surprise since he came from a long line of scavenger birds. Scavengers will eat almost any type of meat, fresh or decaying, and in this sense, they perform a great service in keeping the landscape clean of rotting carcasses. However, the question arises of whether Spirit's diet was representative of what an average crow eats. Well, in some ways, it certainly was representative insofar as Spirit's enjoyment of meats, bread soaked in water, fruits and some vegetables (especially cucumbers) is shared by all crows. However, his diet differed insofar as Spirit did not have to use his scavenger instincts for survival after he was injured. He was handed his food and did not have to go in search of decaying animal matter; he received fresh food, perhaps more rare in other crows' diets. Spirit had a wide variety of foods, which crows on the wing do not often have.

8

A TROUBLING EXPERIENCE

It is especially difficult to write, even in retrospect, about experiences that adversely affect a companion close to your heart, whether human or nonhuman. I find myself reliving the event as I begin to describe an assault that almost took Spirit's life.

The incident happened in June 1991, during the heavy darkness of a new moon. Spirit was in his outdoor living quarters, which had an appearance similar to a chicken coop. It was approximately 14 ft. long by 7 ft. wide and 5 ft. high. The walls and roof of a section of the pen were covered with boards, giving protection from storms. I kept two basins of water and a food dish in this area. As well, this section contained a converted doghouse for extra protection. The floor of the doghouse was covered with layers of sawdust and straw. The remainder of the living area was enclosed with chicken wire as was the door itself. On the ground were numerous pine cones and sticks. Also, a section of the ground was covered in straw, and there was a small spruce tree in a corner of the pen.

The attacker may have been a curious raccoon, a stray dog, or a porcupine, all of which would have been attracted to the odds and ends of dry meat lying about the pen. Whatever it was, I am certain the animal was at least the size of a raccoon because a wide section of wire was indented and pushed forward. This would have taken a great deal of pressure, certainly much more than a weasel or some other smaller animal could have applied.

Spirit must have been terrified by the experience. When I found him the following morning, my mouth fell; I couldn't believe the scene. He was hanging in a perilous position, with his feet and a wing entangled in the wire. He may have been in that position for hours. In his fear and desperate effort to escape the intruder, he must have taken flight and hit the chicken wire, which formed part of the siding of his living space. I hadn't heard any commotion during the night, despite the fact that Spirit's living space was almost adjacent to my cabin. There are several explanations as to why I didn't hear anything. First, I was most likely asleep during the attack. Second, the wind was blowing, which might have smothered sounds. Third, crows are normally quiet during the night; it is conceivable that a single crow would not caw loudly in the night, even if attacked.

I worked quickly to free him from the wire, examining his feet and wings in the process. He was very quiet, probably suffering from mild shock and exhaustion. His wing looked fine, but I suspected his leg was broken, as it was quite weak and dangled limply. I was devastated by his condition, by what could have happened. I felt a sad, desperate anger and frustration, because I hadn't discovered his plight earlier. Yet, there was a positive side to the incident—for the first time I realized how deeply I appreciated having Spirit in my life. Seeing him in this predicament made me realize how precious his life was to me. I realized how quickly events could turn, how easily he could die. I decided that I must double my efforts to enjoy each moment of his company and to learn about his crow nature—there was no time to waste.

Often, it takes a traumatic experience to make us fully realize how wonderful and how transitory life is. Oh,

I always appreciated the special being Spirit was, but I hadn't fully realized until then the extent to which his life touched mine. It is difficult to explain this caring, this appreciation. It is easy to explain to people that you are caring for an injured or disabled bird, which is vulnerable and unable to fend for itself, but it is much more difficult to describe the emotional and psychic rapport that may develop over time between a human and a crow, or any other creature for that matter. Perhaps it is because so many of us underestimate the sentience and intelligence of animals, but when one is open to what a creature such as a crow can teach us, the potential relationship deepens immeasurably.

I have often read reports or listened to stories about how difficult situations, accidents, or troubling experiences lead to new insights and learning on a personal level. Well, it certainly happened to me on this occasion of nearly losing Spirit. I not only had a new appreciation for him and felt a greater kinship with him, but I also witnessed the amazing recuperative power of the crow. Immediately after the incident, Spirit became reclusive and withdrawn and ceased vocalizing of any kind. What a difference from his regular healthy state! This change in him was especially noticeable in the mornings, Spirit's most vocal period of the day. His behaviour had changed completely; he remained hidden away in the converted doghouse, preferring the darkness of its interior to other areas of his outdoor living space. Despite his need for privacy, I always had food and water nearby in case he decided he needed them.

At first, I thought Spirit would die, either from shock or from possible internal injuries that could have gone undetected. I was very worried about him. What I eventually realized was that his quiet, reclusive behaviour was a natural response to the healing process. He only required time and rest—the innate healing capabilities of his body did everything else.

I will never forget the relief and happiness I experienced the day I heard Spirit stirring inside his converted doghouse. His activity was slight at first; it included only small movements of his feet and occasional pecking. I found this encouraging, and for the first time I knew he was over the worst of his ordeal. Later, when he moved to the entrance of his house and looked around, I was jubilant; I could have stood on my head! The following day, he cawed for the first time in over a week. After that his condition improved rapidly, and soon he was much more active, moving about in his usual vigorous fashion. It seemed as though we both had been given another chance.

9

CONTEMPLATIONS IN FOUR PARTS

I

In the spring of 1993, several crows and a pair of ravens tended the general area where Spirit lived outside my cabin. One of the crows, probably a female, seemed very interested in Spirit and frequented a nearby tree from where she could watch his pen. She often cawed, but he never replied, seeming to prefer a silent, guarded attitude towards her.

Later that summer, the crows behaved less cautiously towards me. I gave them scraps of bread, which I was unable to use or had bought especially for them. I had begun feeding the crows the previous winter during a severe cold snap. At one point I was feeding four crows, a pair of ravens, a seagull, numerous blue jays, and a crowd of other, smaller birds.

In a letter, Elizabeth Turner of Victoria County, Nova Scotia, wrote that she, too, feeds crows:

> *My husband made a feeding tray and placed it on a post in front of the kitchen window. It attracts many varieties of birds, especially crows. As soon as daybreak we are awakened by 'caw-a-caw.' I save scraps of food and mix them with flour to form dumplings. Then I place them on the tray, and usually one crow is on duty in the highest tree, watching me and the tray. As soon as the food is placed, the crow gives a shout and soon a flock appears. They gorge themselves then carry away the balance, which I assume is cached for future use. This goes on all year. They are very clean and tidy, and carry away anything they can't eat, such as bones. Often those are found at the base of trees further back in the woods.*

In principle, I disapprove of feeding birds, believing that it may make them dependent on humans. But it is certainly good to assist birds during severe cold periods, in the depths of winter, or in other emergency situations. I must confess to breaking my principles, though; like Elizabeth Turner, I thoroughly enjoy their close companionship year-round.

Often, this close companionship can result in interesting experiences. For example, one afternoon in late April 1993, I happened to look out my window at the right moment to see a huge raven sitting in an apple tree to the rear of my cabin. It was probably one of the ravens that had been frequenting the area for the past several years. To my surprise, the raven suddenly flew from the tree and landed near Spirit's pen. It hopped directly to the pen and peered at Spirit through the wire siding. There was certainly no hostility involved in the raven's behaviour. Rather, it appeared to be curious as much as anything, and flew off only after noticing my movement in the cabin window. Spirit, for his part, took everything in stride. He simply remained in a sitting position, and watched the raven. This was a memorable experience for me because it was the first time that another bird had expressed such curiosity about Spirit.

II

Spirit seemed to be capable of distinguishing between colours and sizes, and to a lesser extent shapes, because he responded differently to variations in these things. Spirit was alarmed if I wore clothing which was new or unfamiliar to him. My theory is that he was alarmed by colours or combinations of vivid colours he hadn't seen before. For example, if I suddenly appeared in a bright yellow shirt, he was apt to be defensive. To a certain extent, colours seemed more important than shapes to his distinction of familiar and unfamiliar things in his world. Of course, the size of objects was also important because, as you might imagine, large objects tended to scare him, although he was sometimes able to adjust to their presence in his environment.

One evening I approached Spirit wearing a red, blue, and white checked shirt. As I hadn't worn this shirt before in his company, he became quite alarmed and fled from me. I thought the sound of my voice might soothe him, but it wasn't enough to quell his fear. I was disappointed. Any pride I had felt in Spirit recognizing me as an individual was thoroughly deflated! Perhaps he was identifying me only by my clothing, rather than something more personal, like the shape of my nose or the sound of my voice. I suddenly felt that I didn't understand how crows identified "friends" or familiar objects. I tried to be open-minded about the whole thing and attributed my disappointment to cross-species misunderstanding. I liked that—it sounded scientific. If colour was the factor crows used to identify friends or familiar things, there was nothing I could do about it. I might as well accept it as a fact of crow consciousness and not take it so personally.

However, I wasn't content with this theory. I knew something was missing; there was an element I had overlooked. I pondered Spirit's reaction to my brightly coloured clothing. The whole truth of the matter occurred to me suddenly, and it was somewhat opposite to my initial perception. To Spirit I was certainly much more than a red checked shirt. In looking for complicated answers, I overlooked the obvious—the brightly coloured shirt I wore on that occasion had been totally foreign to Spirit and was therefore something to be feared. He recognized me, I reasoned, but that predominantly bright red "thing" was something else. The fact that I was wearing it didn't make it any less frightening to him. The red shirt was a barrier between myself and Spirit. I figured the shirt was like a fox between two rabbits, preventing normal everyday interaction between friends.

III

One early spring morning in late April 1994, I was awakened by loud cawing from Spirit, who was having a great time greeting the birth of a new day. There had been a heavy frost, and this may have affected his mood. As I lay in bed listening to his loud singing, I slowly drifted into that almost magic state between sleep and wakefulness. I found myself imagining that crows everywhere were cawing in unison. Behind my closed eyes, I could see thousands of crows flying about in one grand, harmonious pattern. Startled by the vivid scene, I bolted upright in bed, wondering if there really is an internal or instinctive sense that psychically connects the members of a particular species. Do crows share a certain internal harmony, which can synchronise their actions? I understood that in my dream, each crow in the sky knew what the others were doing and was instantly aware of changes in movement and flight pattern. Of course, this was a dream, and I have no proof of such profound synchronicity. However, I have observed how well a flock of crows are able to synchronize their movements. When a lead crow changes direction, the others quickly follow its example.

Later that day, I took a walk along the eastern shore of Minamkeak Lake. At one point along this shore, there is an old, partly dead hemlock, its limbs extending antlerlike over the water. As I stood by the tree, I remembered that I had once noticed several crows perched on its limbs. I recalled how my foot slipped while watching them, causing loud clapping sounds to echo over the water as loose rocks came together. The crows sharply cawed a warning as they flew away to the southwest. I was able to watch them dive and sweep towards a grove of pine trees on the distant horizon. This memory prompted me to think of the life Spirit might have had and of the accident which tragically altered his destiny. In the midst of recollecting, I smiled as my thoughts shifted to the experiences he and I were sharing.

Upon returning from my walk at Minamkeak, I decided to take a late breakfast at the Blarney Stone restaurant in Hebbs Cross. As I sat by a window eating my toast and enjoying the late morning sun, I saw three crows flying in the direction of Fancy Lake. They were making playful movements in the sky, circling, diving, and generally behaving in a manner characteristic of crows in the spring of the year. I watched them, oblivious to my breakfast and the other customers in the restaurant. I wondered whether such awe inspiring moments as this prompted people like Francis Allen and Millicent Flicken to write about the playful, aesthetic nature of crows and other corvids.

IV

As a young boy, I admired crows and ravens. At the time, I remember noting that people were always critical of them; they didn't have a good thing to say. So, partly for that reason, I decided I loved crows. Besides, I was rather rebellious and always favoured the underdog. To most of my friends and neighbours, crows were simply scavengers that wreaked havoc in everyone's garbage or in their gardens and committed a variety of other unsavoury deeds. They were always getting cast as destructive birds. I can't recall anyone ever describing crows as beautiful, unlike blue jays or other birds of striking colour. The crow, while recognized for its intelligence and deviousness, was largely ignored or cursed.

Even as a child, I realized there were lessons to be learned from crows and ravens. When I saw a crow flying, I would sometimes think, "I wonder how they see the world. They are so intelligent, surely they see it in an interesting way." I realized early on that there were many things people could learn from two-winged beings, if we were patient enough to watch, listen, and admit that another species could teach us important life lessons. Years later, the time spent with Spirit convinced me more than ever of the value of living closely with other creatures and being open to learning from the shared experience.

Crows are social birds and have probably always lived close to human society. Humans have had ample opportunity to witness crow behaviour and to enshrine it in legend and myth. I like to believe that those legends and myths of Native North American peoples were, in part, created to express admiration and respect for the intelligent and cunning nature of the crow, genetically preserved and expressed in the actions, tendencies, and behaviour of each generation of crows.

If we examine the cultural traditions across North America, we will find that Native peoples have a long-standing recognition of the value of crows, along with everything that lives. Today, it is important that all humans share that recognition and move from the isolation that places us apart from everything else on this planet. Our alienation is often fostered by the technological world in which we live. We forget we have a kinship with other creatures and the urgency to share the earth in a responsible manner.

10

LIVING LEGENDS: What Crows Teach Us

Leipsigaek Lake, near Bridgewater, Nova Scotia, is a source of refuge, a place I go to gain balance and strength. I have received much inspiration while hiking its rocky shoreline, or exploring a secluded cove, or gazing over its waters. The last time I went there, I visited a small inlet and a rock fireplace I had constructed several years ago. The site is flooded for most of the year, accessible when the water level recedes in late summer or early fall. This fireplace is special, as it was here I prepared sacred eagle feathers, praying deeply for direction and guidance in my efforts. Those feathers are symbolic of the spirit world and of prayer and spiritual power, among other things. Looking back, I understand that the process was an initiation as, intuitively, I knew all the proper steps, even though I had never prepared feathers before. I will always remember that important event in my life; the sacred fire will always burn, and whenever I need the flame it becomes evident in my life as a source of inspiration and guidance.

It was at Leipsigaek in 1990, near this fire site, that I first truly learned from crows. The insight I gained then preceded some of the more meaningful things I would learn from Spirit. I was relaxing, surveying the waves and the distant shoreline of Leipsigaek, when I heard the loud cawing of several crows on pine trees further inland. I suspected the crows had been observing me for quite some time before I was alerted to their presence. As I watched those crows, I sensed that they could teach me something. I felt a deep love for those birds at that moment. The way they sat on those trees, patiently observing my actions, silently watching my movements, made me realize that if I could demonstrate such patience and silence in my own life, I would more easily reach my goals. My life would be much more relaxed and at peace. At the same time, I felt embarrassed by the fact that I had lived with Spirit for approximately a year, and I had failed to sense those things from him. In retrospect, I now realize that I was so involved in the daily work of caring for him and of meeting his living requirements, that I failed to notice some of his more subtle crow traits.

Of course, it is difficult for us to cultivate the patience of the crow. We are accustom to living our lives motivated by desires and whims. Patience is cultivated with much practice, including the observation of creatures that exhibit patience in their behaviour. I now try to foster patience in my own life on a daily basis, to prevent costly mistakes or hasty decisions and to maintain a sense of peaceful balance. It doesn't always work; there are times when frustration overcomes me, and I'm less able to endure or change to meet a given situation.

I did eventually realize from Spirit that patience and endurance are inseparable qualities. Just consider the trials he had to endure in his life—the injuries, the adjustments to new environments, and adapting to the proximity of humans. Perhaps the most important lesson I learned from Spirit is to be more observant of my environment. We take so much for granted around us and usually observe only those things we expect to see. Yet, Spirit seemed to observe even minor things or small changes in his environment. For example, he was quick to notice the movements of insects, the swaying of grasses, the movements of other birds, and the sound of my car as distinct from other cars. As well, his keen eyesight could easily recognize my presence at a distance of 75 yds. to 100 yds.

Ravens are incredible in this respect, having highly refined senses. I learned this through personal experience. Last year I tried several times to observe a particular raven with my binoculars. While I succeeded on several occasions, there were many times when the raven was alerted to my presence and intention. Once, I tried to observe it from the loft of my cabin where I had been sleeping. On this particular morning I awoke to the loud voice of the raven that was sitting on a tree about 40 yds. from my window. Lying in bed, I carefully reached for the binoculars, putting them to my eyes. To my amazement the raven flew away—it must have observed my movements with its keen sight! Little wonder ravens and crows have survived through the generations; their marvellous sensory awareness has assured their continuity.

II

If you are more than a little curious about crows, you will want to observe them as often as possible. It will mean outdoor adventure, visiting a number of environments to study the crow's reaction to various circumstances or conditions. As a matter of course, whether you realize it, you will become absorbed by crows and may even acquire certain of their characteristics, which most attract you. This can be somewhat like a shamanic transformation, a personal experience of spiritual significance. You may smile in disbelief, though I can assure you this is likely to occur; crows are mesmerizing. Their behaviour is so interesting and so obviously guided by intelligence, that it captivates the observer. In any case, you will gain a heightened appreciation of crow behaviour and become more willing to learn from other species and to apply this learning in your own life.

Dorothy Knowles, of East Hampton, New York, wrote to me about the significance of her experiences with crows and ravens. She remarked:

It happened during my 1993 vacation to the southwest. One morning at Lake Powell I went for a walk. It was cold and brisk, and I thoroughly enjoyed being alone with my thoughts. Off in the distance, in back of the lodge, was a huge flock of crows interested in what was thrown out from the kitchen. I smiled and went on my way at a quicker pace. Seconds later I heard this incredible sound moving just over my head. I looked up to see this crow just about scaling my head. The sound was his wing beats—powerful, rhythmic. All I could do was follow him with my eyes and exclaim, 'Wow!' over and over again. It was all I thought about the whole morning on the tour bus. I wanted to remember that sound forever.

After returning home, Dorothy discussed her experience with a friend and realized the spiritual importance of her encounter with the crow. Later, she tried a visualization exercise, which she learned from Dr. Michael Samuels's book *Healing With the Mind's Eye*. She wrote:

I pictured myself back in Sedona, climbing over the rocks and finding a cave. In the cave was a hole going down, down, down. When I slid through it to follow its direction, I came out at the bottom of a canyon. Ahead of me was a raven (a member of the crow family), sitting on a ledge. I looked to my right and noticed another cave. The raven flew over to it, and out of the cave came a Native man whom I felt had great wisdom. I climbed up to the cave and stood between the raven and the man. I became a part of both of them and then just the raven. I took off in his form and flew the skies over Sedona. It was fantastic—the freedom, the flight. When we returned I bid goodbye to the wise man and the raven and thanked them for a wonderful journey, ending with, 'I hope we meet again, soon.' With that I fell asleep.

Next morning, at the office, I walked over to the window ledge where we place outgoing mail. I looked out the window and saw a group of six crows. One broke away and began walking up the incline towards me. At that moment a co-worker passed by and said, 'Hey, look at that crow!' My words from the night before came to mind, 'I hope we meet again soon.'

Since those early encounters with crows, Dorothy has had many experiences. She learned that her totem animal is the raven, and that she has a spiritual guide named "Swift Raven." She remarked that "crows always seem to be there for me. They are a 'sign,' which I always find comforting." In fact, she began feeding the crows that frequent the area where she works. Later, she was rewarded with the gift of a crow feather.

Dorothy has kept a written account of her crow stories that reveal what she has learned from them, a couple of which I include below.

Last Christmas, one of our sons was leaving with his family after a lovely week's visit. A five-hour trip was ahead of them. As they drove away, I asked that they have a safe trip, to be watched over and protected. At that moment two crows flew out from somewhere, passed over their car, and flew in the direction they were headed. I knew for sure they would arrive home safely. There was that 'sign.'

Another sign appeared last winter, when I went to the hospital for a breast biopsy. For weeks my mind whirled over the most negative thoughts; I was a wreck. Leaving for the hospital in the early hours of the morning, I opened the door and stepped into the dark. A crow flew overhead. Part of me knew this was a good sign, but the other part was too scared to think positively. I was about to have another mammogram so the doctors could pinpoint the section to be removed. After three pictures, they said they couldn't find the area they had previously been suspicious of; the operation wouldn't be taking place. Back in my room I ate up the feeling of relief.

While I waited for my husband to come back to the hospital to take me home, I sat looking out at the beautiful tree pressed up against the window. Two crows flew into the tree, cawing and puffing out their chest feathers (I had never seen that before). What a performance. They seemed to be just as elated as I was. Walking down the hallway, I looked into the rooms as we passed. None of them had a view of a tree.

III

Of course, interacting with the natural environment is beneficial and therapeutic on numerous levels. There is a wonderful release from stress and a feeling of relaxation to be gained from nature—from casually exploring lakes and bogs, to feeling the moss under your feet as you walk in a shaded conifer forest, or to hearing the sound of crunching leaves underfoot while walking in a hardwood stand in autumn.

More recently I find that, as I explore the natural environment, I rely on my intuition to guide me to interesting experiences. One afternoon in July 1992, I suddenly felt a strong urge to take my binoculars and visit Minamkeak Lake. Walking directly there, I was casually strolling over rocks, when I heard crows cawing behind me. Turning sharply, I saw four crows flying in a northeasterly direction to settle on pine trees in the distance. Focusing the binoculars, I was delighted to see two crows fly from the trees and land on rocks near the water. I enjoyed observing how they moved about on rocks, heads bobbing, looking from side to side, and how they walked slowly, then hopped onto another rock. I watched one of the crows briefly put its head in the water. I was happy that I followed my hunch to visit Minamkeak that summer afternoon.

On another occasion, I visited the "barrens" near Leipsigaek Lake. The area is relatively flat in most places, with rocks and large boulders scattered about. It was mined for gold from about 1890 to 1950. I went there on a hunch I had earlier that morning. The sky was partly overcast with a strong wind from the northwest. I drove my car over the old mining road that winds through the barrens and parked near a bridge over still water, before proceeding on foot to a large rock outcropping nearby. As I made my way to the formation, I was struck by the variety of plant life—wild pear bushes, blueberry, huckleberry, and bunchberry plants grow in abundance there. I noted numerous clumps of sweet gale, smelling as delightful as the bayberry, and large clusters of lambkill, along with scattered samples of leather leaf, ground juniper, and wild sarsaparilla. There were patches of crowberry covering the approach to the rock outcropping and over the formation itself.

From the top of the outcropping, I could see a considerable distance over the barrens. With my binoculars, I could clearly locate the still water brook at 200 yds., meandering through the barren landscape. When I looked to the northeast, my view was mostly the tops of young birch and poplar trees, their branches swayed in the breeze and their leaves shimmered in the soft evening light, creating an interesting wavelike effect, which continued to the horizon. For several moments I watched as a crow flew near the tops of the trees, disappearing in the heavy growth of leaves and branches. Later it reappeared for an instant before vanishing in the cool shade of the foliage.

As I returned home that evening, I saw another crow flying south, its wing movements steady and untiring as it made its way over a long cove. I was reminded of the determination of the crow, a quality so exemplified by Spirit in his propensity for overcoming the handicaps in his life and adapting to his limitations—or possibilities. I observed determination in his pecking behaviour—he would not be deterred until he had finished whatever he was doing with his beak. It was equally difficult to dissuade him from taking a thorough bath. Spirit was adamant about keeping his living space clean of feathers, and he did so, no matter how often I replaced them. This kind of persistence is indicative of the crow's nature—it does not compromise its natural talents. If we humans could follow our path in life with the same tenacity towards overcoming obstacles, we would surely succeed in reaching the goals we set. We would have learned well from the two-winged ones.

IV

There is little wonder that crows are very often the subjects of legends, folktales, and storytelling traditions around the world. I make those remarks in light of the five years I shared with Spirit. He had keen sight and hearing, and his other senses were no less acute. This was most evident on those occasions that involved physical contact. I have mentioned how crows thrive on touch and grooming behaviour. Spirit could be very gentle, using his beak with a sensitivity that we tend to associate with our hands. He would take my finger in his beak and apply pressure ever so gently while I massaged his head. When I stuck one of my fingers through the chicken wire of his living area, he would often latch onto it, gradually increasing pressure with his beak. But Spirit was very sensitive to the amount of pressure he was applying. Often when I approached him, he would hop towards me, making various low-pitched crow sounds, which told me he wanted affection. Even his manner of hopping—rapidly with wing movements—signalled a desire for affection. It is this kind of sensitivity which makes crows and other corvids legendary birds. The crow stories I have collected over the years exemplify the intelligence and unique character of these birds.

Marilyn Burns of Yarmouth, Nova Scotia, related the comic intervention of a crow in a quarrel between two pheasants:

As I glanced out of my dining-room window one day, I saw two pheasants going head to head in an altercation. Hopping excitedly from side to side over the pheasants was a crow. He was so like a referee working a boxing match, I had to laugh. Finally one pheasant dominated and the other ran away. The crow flew to a nearby tree and from the topmost branch cawed the news for at least five minutes.

Phyllis Gillis, from Orangeville, Ontario, recounted a story, illustrating both the intelligence and determination of crows.

Last June, my son decided to remove the hardtop from his jeep and replace it with a summer canvas top. As there was no place to store it, he left it in the backyard. The hardtop has windows on three sides; the part that slides over the windshield is open.

A few days later, two curious crows came to investigate this apparition. Looking through the windows, two crows—their reflections—stared back at them. How excited they became, cawing and flapping their wings as they pecked at their images! They kept at it for so long and with such noise and excitement that we thought they would literally have heart attacks. Finally, at dusk, they left.

On day two, they returned, repeating the previous day's performance. This time they had a new plan. After the noisy greeting to their images, they raced inside the jeep top, through the opening. They stopped dead in their tracks when they saw no crows there. You could almost see their confusion. With much cawing and flapping, they raced outside and ran around the top. When they stopped at the windows they could see 'them' again. How excited they were! So back inside they went, and the whole process started again.

On day three, they came up with another plan. They flew quickly to the window—always the same one as the sun didn't give their reflections in the others—and became excited when they saw their 'friends' again. But this time, only one crow raced inside, while the other one stood guard at the window. Again, the one inside the hardtop found it empty. He raced outside to tell his pal. They changed places.

The opening of the hardtop is big enough that we could see what they were doing inside, and their noise always let us know they were there, so we could dart to a window to watch. Bird number two ran in, and after checking all the corners, he went to the window to tell his friend, who then also raced inside. Looking out the window, they must have seen the reflections again as they became very excited, and with a lot of cawing and flapping of wings, they hurried outside. Racing around the top, they came to that window again, where, behold, their friends were inside. They returned to the opening, but no crows were there. I'm sure if they were human, they would have stopped and scratched their heads in bewilderment.

This went on most of the day. At this point we found it not only amusing, but we were concerned for their health. On day four, two obviously tired crows appeared, and this time they had another plan. First they checked to see if their friends were there. Their noisy cawing and wing flapping told us they had been spotted. Instead of going inside, they quickly flew up to the tree above and waited ... and waited. Nothing happened. So back they went to the window, saw their reflections, and flew back to the tree. This waiting game went on most of the day. On day five, they repeated day four, but quickly. As on day four, they checked inside before leaving.

I never saw such intelligence and determination in birds. We were completely fascinated by the whole process. However, our concern for *the welfare of these birds (as well as the hardtop), prompted us to buy a tarpaulin to cover it. When they no longer saw their reflections, they simply flew away.*

The following is a story written by the late Burnice Gilchrist, who had been a resident of Pictou for the last sixty years of her life. She passed away on the 30 August 1995 at the age of ninety-one. Her story was compiled by a family member from several of Burnice's rough drafts. It, too, demonstrates the intelligence of crows.

When I was five years old, my parents, with four children, took up a homestead in Saskatchewan. It was about a mile and a half south of the North Saskatchewan River. Our new house was built on a hill that provided great coasting in the winter. The summer I was ten or eleven my father had built a smokehouse to cure meat. He had cut into the side of the hill, just below the house, and had unearthed a pocket of good clay. All summer we enjoyed making dishes and little creatures out of the clay, baking them in the oven after mother took out the bread.

Early one morning in late summer, just as we were called to breakfast, father glanced out the north window towards the river and called us to tell us two visitors were walking towards our home. As visitors were scarce in those days, we rushed over to see who was coming. We noticed two crows walking through the pasture towards the house. One could fly, and the other was dragging an injured wing. We guessed it had probably hit the telegraph wires when it was learning to fly. (The telegraph wires crossing Canada went across our farm.) Some crows had made their nests in a poplar grove near the wires.

We watched through the window and wondered why they were coming to see us. We soon found out. They stopped at the clay pile, then the older (larger) crow took bits of clay in her beak, wetting it as she did so, and packed a cast on the injured wing of her companion.

Our parents told us to put food out for the crows but to be very careful that we didn't frighten them. You may be sure that they were well fed. We also kept our collie from chasing them. The crows stayed near our home for several days, I don't remember exactly how long. When the clay cast came off, the wing was healed, and both crows flew away. I have often wondered if anyone else ever witnessed a 'Dr. Crow' performing such an operation.

John G. McKay of Amherst, Nova Scotia, related an interesting story about his maternal grandfather, Joe Gothreau, who had an especially clever pet crow named Dick.

Joe Gothreau was an avid hunter, trapper, and guide. During one of his excursions, he acquired a nestling crow, which he was determined to domesticate, much to the chagrin of my grandmother.

The crow, which he named Dick, being naturally intelligent and opportunistic, gradually developed a kinship with the old man, a loyalty unsurpassed by any dog. In effect, given its way, the crow would have been inseparable from the old man. However, my grandmother asserted her right to a peaceful domicile, and to preserve domestic tranquility the crow was confined to the back porch whenever Joe was at home.

Now Joe was a shrewd trader in the manner of the times and was not averse to the occasional wager, particularly when the opportunity arose for fixing the odds himself. It was, after all, a form of trading, a playing upon the same sense of avaricious hope essential in all forms of barter. In that regard, Joe had one trick that not only proved lucrative from time to time, it never failed to earn him the accolades we all inwardly or openly crave.

In warm weather, Joe would often gather with his friends at

Victoria Square, Amherst, to gossip and discuss the latest news. Prior to leaving for the gathering in the square, Joe would instruct my grandmother: 'Let Dick out a half-hour after I leave the house.' No request pleased my grandmother more.

Joe would slowly make his way to the square, in plenty of time to arrange the dodge, which could only be accomplished if there was a stranger among the loungers on the benches. The old man would select his mark, manoeuvre himself into a strategic position on the appropriate bench, and wait.

When the half-hour was up, my grandmother would release the crow, with good riddance. Dick was also a pawn in the game, but one who participated eagerly. Having but one concern—finding the old man—the crow would begin a long spiraling climb, rising high above the town, to scan the streets and avenues far below for his soul mate and benefactor. The longer it took him, the higher he climbed, until finally, his keen eye would spot his quarry and he would begin his descent.

Regardless of the stranger's response—for none of Joe's knowing cronies would take the bait—Joe would reply, 'I've got fifty cents that says I can have that crow sitting on my foot inside of five minutes.'

Of course, only a fool would pass up such ridiculous odds, and the stranger, being the freshest, least tarnished fool among them, would promptly cover the bet. 'The man is crazy,' the stranger would affirm among the loungers.

Joe would then cup his hands to his mouth and give out with a series of the greatest blats of nonsensical yodelling and caterwauling ever aimed skyward. None of this meant a thing to the crow. If he hadn't already spotted the old man, he used the familiar noise as a beacon on which to ascend, arrow-straight, toward Joe, who would be sitting cross-legged on the bench, and land on his foot. The stranger, dumbfounded by the experience, would eventually mutter something to the effect that the show was worth five times the price of admission.

Several people wrote to me describing the many wonderful and often unusual experiences resulting from having a pet crow. Linda Goodin of Upper Clements, Nova Scotia, told the following story of a pet crow that became very attached to her family.

After moving from British Columbia in 1977 to a cabin in the woods of Annapolis County, Nova Scotia, I often heard my husband, George, recalling boyhood memories of his pet crow Timmy. His tales of wonder and happiness together with his crow absolutely amazed me and our then small son. We often talked about getting a baby crow to enjoy the experiences George had and the amazing relationship between country boy and crow.

About 1983, when our son Josey was five years old, someone in the area had mentioned having a baby crow, which had been taken from its nest. George pursued the matter and soon afterwards came home with this baby bird whom to this day we often think of and talk about.

We named the crow Dakota, which soon became 'Dakcrowta,' and it not only became our guardian but also had a wonderful relationship with our dog, Cochise. None of us had ever known just how intelligent a crow could be, and as he grew he became totally dedicated to all of us.

Whenever I put laundry on the line, there was Dakota unpinning every last piece of clothing and jumping up and down, laughing at the fun in this. He would wait for us to play badminton, perch right in the middle of the net on the lawn, and every time the birdie fell on the ground, he'd grab it. He'd jump up and down with the birdie in his mouth, then fly onto the roof, run across to the other side of the cabin, and drop it so that we'd have to run around looking for it! Whenever we went for walks up our country road, he would literally walk with us. He'd never fly, but he might grab a ride on our dog's back. He would also ride the handle bars of our son's bicycle.

One of the most amazing things he would do was listen for the sound of our vehicle whenever we were coming home from town. He would fly the half mile or so to meet us and then fly in all the way, hovering over the roof of the jeep. Whenever he was bad (and knew it), he would fly onto my shoulder and stick his head into my armpit as if to ask for forgiveness. He was utterly wonderful!

Then came the times when we had company. Unfortunately, he became very possessive of us. He would corner people in the outhouse, wait until they were busy, and start pecking at their toes! This was rather funny, until he made someone's feet bleed. He also cornered some kids against a tree a couple of times and scared them half to death. He was actually being protective of our son. We started thinking he was turning into a sentinel after a young boy held up his fist to him, and the crow immediately flew at his head, just enough to scare him and ruffle his hair.

We finally decided to let him go and took him back to a lake a few miles away. After letting him out of the jeep, we drove away, only to look back and see him flying after us. It was heart wrenching. A strange vehicle came up the road then, which seemed to scare him, and he flew back towards the lake.

A few days later I was so upset over not having him around that we went back to get him, but he was gone. Hopefully he had or is still having a wonderful life.

Sheila Wilson of Pugwash, Nova Scotia, related some rather unusual experiences with Joe the crow, whose intelligence and sense of adventure won him a good deal of trouble.

Joe came to live with us on a summer day in 1957. He took up temporary residence in a cardboard box in the back porch, but it wasn't long before he had the run of the house. He thrived. Soon he was big enough to learn to fly. We brought him out on the lawn and took turns sort of tossing him in the air. After many crash landings, he finally got the hang of it, and we all felt like proud parents witnessing our baby's first steps.

He followed us around like a dog. He came to school and waited for us in a large tree on the playground. He followed us home again, flying just above our heads all the way, occasionally landing on our shoulders. When we went for a row up the creek in the dory, he followed us, soaring high over our heads or perching on a dead tree on the bank of the river.

Everyone for miles around knew Joe. He would sometimes venture off on his own. Going to town was usually a disaster. Mother would get calls from local merchants, saying that Joe was harassing people on the street, landing on their shoulders, begging for treats, or stealing items from cars if they were foolish enough to leave their windows open. Men's hats were fair game, especially the ones with feathers in the band. Mother would walk to town to get him. Joe would see her coming and fly down to light on one of her shoulders for a ride home.

Our next door neighbours were an elderly couple, and Joe visited them often, usually early in the morning just after the milkman had made his delivery. You see, Joe liked cream. In those days milk came in bottles with a cardboard cover on top. The milk was raw not homogenized, and when the milk sat for a while the cream rose to the top. Joe would skilfully remove the cardboard top, drink the cream, and replace the top. The neighbours couldn't figure out why the milk bottle was only three-quarters full every morning, until they caught Joe in the act.

Joe lived with us for two years. One late August morning he went out as usual but did not return that night or the next. We watched for him the rest of that summer and fall. Joe didn't return to his box in the porch, but for several summers after that one, whenever we went for a row up the creek, a large, glossy, black crow would soar over our heads and follow us a ways. We never knew whether it was Joe, but we like to believe that it was.

I have heard that if a person is lost in the forest, they should call upon the crow for help. If you call the crow, you should watch carefully for it to appear because it is certain to happen. It will fly near you or pass directly overhead. When it appears, you should follow the crow, even though it is a trickster and may make you walk through challenging terrain; the crow likes to joke around. However, in the end, it will bring you safely from the forest to a familiar location.

My long-time friend, Peggi Thayer, who presently lives at Pictou Landing, is very much aware of the crow's intelligent nature. She told me a story about asking a crow for directions. The incident happened several years ago in Windsor, Nova Scotia. In Peggi's words:

I got up early and headed out walking to the hospital to visit a friend, Harry, taking what I thought would be a shorter way than walking the highway. After about a mile and a half, I realized that the road I was on was going off to the right and away from where the hospital should be. About that time I heard a flutter in a tree over my head, and a crow called out as it fluffed its feathers and settled down on the branch. I looked up and said, 'Good morning crow,' and walked on a few steps. Then I started thinking of what I'd been told about crows—that they were messengers [the means for receiving guidance, inspiration, and communication from the spirit world], could be tricksters, and could give directions. I thought, 'Well, no harm in asking.' So I went back a few steps, looked up and asked the crow which way it was to the hospital. Sure enough, the crow let out a 'caaw' and flew off to the left. When I watched his direction, I saw that it led across a railroad track and a big cow pasture to the hospital building a mile or so away, on the other side of the field. In that field were a few head of cattle, including a bull.

Well, the cows were headed the opposite direction I needed to go, so I decided to carry on, keeping a close eye on that bull. I got to within a hundred yards of the far fence, when I saw the ditch and a marshy area on the other side of it. I found a fairly dry place to cross the ditch, but there was no way to avoid the marshy spot. I wasn't about to turn around since the hospital was only on the other side of the fence and across the road. So I slogged through—good thing I was wearing boots—and arrived at the other fence, mud splattered to my knees, only to find that crow, perched on the fence post, making a noise that for all the world sounded like he was laughing at me! I had to laugh too, as I said, 'Well, thank you anyway!'

I wiped off as much mud as I could on the hospital lawn and had a great visit with Harry. It was the last time I saw him.

I also knew Harry, who was a good friend and a special person. Born and raised in Windsor, Nova Scotia, he had Native ancestry, which he said was Kootch. He knew much about the ways of birds and animals, and I think that perhaps he was particularly fond of crows. He certainly travelled the woods a great deal in his younger years and must have had countless opportunities to observe crows. He told this story, which suggests something about a sense of community among crows and about Harry's spiritual well being.

You know, when you spend a lot of time in the woods, you see the strangest things. One time I was out the woods road when I heard a lot of flutterin' and goin' on. I turned toward the sound, and pretty soon I came onto a little clearing, and in that clearing was a bunch of crows settin' around on the ground. What I noticed next was that they was settin' around pretty much in a circle, and in the centre of that circle was a dead crow!

They kept settin' there, quiet like, and once in awhile they would start up mutterin', you know, just makin' this odd, soft little mutterin' sound, and that sound went all around the circle of 'em. I watched for quite a little while. Looked altogether like a funeral. After a bit, I looked away, just for a couple of seconds, when I heard a loud, sharp call go out. There was a great flutter of wings, and all of 'em took off right sudden like. And you know, when I looked around, that dead crow was gone too.

People like Harry were special because they travelled the forests so much, gleaning practical knowledge about how to live independently, without always having to rely on others for their survival. Just as important was Harry's sense of wonder, his faith, and his sensitivity. It was those qualities that gave Harry the inspiration to roam about the forests and rewarded him with the rare gift of a crow circle. In some cultures, for example, among traditional Algonquin peoples, this would be considered an important event in one's life. It is a rare medicine gift, indicating a person has developed a degree of spiritual power.

Unfortunately, in mainstream Western culture, we have lost this sense of the "sacred." So often we do not notice or recognize the many natural gifts with which life blesses us. Spirit endeavored to reinforce and to make me more aware of that sacredness. On one occasion I saw Spirit in a vivid dream. I remember waking in the morning, thinking about the dream, and wondering what it meant. Upon visiting Spirit later that morning, I found the most beautiful feather laying on the ground near him. It was a special gift, which he gave to me in the dream. Our lives will present us with such gifts if we are alive and alert to the possibility of them.

11

THE CLOSING

The winter of 1994 proceeded in the usual manner, with Spirit pecking a multitude of holes in his cardboard house, showing his usual behaviour towards my daytime telephone conversations, splashing vigorously in his water basin, and displaying his usual fascination with my painting techniques. However, February included some unusual events, to say the least.

I returned home one evening to find Spirit sitting on a table amongst my oil paints and brushes. This was strange and puzzling to me because he appeared to be very excited and somewhat nervous. After he calmed down, I returned him to his living space and retired for the night. At about three o'clock in the morning I awoke to noises of the loud flapping of wings and objects hitting the floor. I quickly jumped from bed and hurried downstairs to find the floor strewn with papers and paint brushes. Spirit was sitting on a chair in a corner of the room.

I couldn't understand what was so agitating him. I fell into a chair of my own, trying to figure out the possible reasons for the events of the past few hours. I was coming up short of an answer, when from the corner of my eye, I noticed movement. It was only for a brief second, but it was definitely movement, and it was coming from behind some paintings leaning against a wall. I sat still, hardly breathing. I watched, the way a spider watches its web from the perimeter, or a crow watches the landscape from a tree top. Finally, there it was again, the same movement. I was astonished to see a weasel peering from behind a landscape painting! For an instant our eyes met, then in a flash, the weasel ran to the steps and bolted upstairs.

I was stunned. To find a weasel in my cabin was both remarkable and, at the same time, my worst nightmare! The weasel was beautiful, with its snow white fur and black-tipped tail. It was 12 in. to 15 in. long. I found it remarkable to be sharing such a small space with a weasel and a crow simultaneously! The nightmare, of course, was that a weasel and a crow were a deadly combination. I knew the weasel might easily kill Spirit for they are smart predators that can strike quickly and viciously.

I immediately set to work reinforcing Spirit's living area, securing the walls and making them much higher. I believed I had a chance to keep the weasel away from Spirit. I noticed it had raided Spirit's food supply—his living area and food dish were completely empty of the usual scraps of food.

For the following week, the three of us shared my cabin. I found it was next to impossible to keep the weasel from invading Spirit's space. No matter how much I tried to prevent it, the weasel always seemed to reach Spirit's food. In the end, I may have won a partial victory, but I was never quite sure. I found it was more satisfactory to feed the weasel. I always made sure there was meat lying about my cabin floor. This appeared to be the best strategy since Spirit was able to live a more normal, contented life.

I was in the cabin loft one night, sitting on my mattress as I prepared for bed, when I noticed the weasel watching me. It was standing between two boxes about 4 ft. from my bed, which was simply a covered mattress on the floor. I realized then the weasel had been making its home in the cardboard boxes where I kept my belongings, and that we had been sharing the loft for several nights. As I sat there, the weasel edged closer and seemed to be extremely curious about my feet. It was moving cautiously but inched its way very close to my feet before retreating to the protection of the boxes. It was really quite funny to watch the weasel furtively approach my feet.

I enjoyed the adventure of sharing space with a weasel and a crow. It was like living a folktale without being certain of its ending. However, I was more concerned about Spirit's safety. I realized I was courting disaster, and that I had to take steps to protect Spirit. I decided to livetrap the weasel in a box. To build the trap, I placed a 2 ft. square piece of masonite on the floor, near the steps to the loft. I then placed a cardboard box (about 15 in. x 20 in.) on this base, lifting one side 5 in. or 6 in. off the masonite with a small stick, to which I had tied a long string leading upstairs. I placed a weight on top of the box so that it would fall faster and securely trap the weasel. Inside the box, I placed bologna and hamburger.

I waited for the weasel to take the bait. It wasn't long before I noticed it peeking downstairs from the top step. Finally, from my chair near the woodstove, I saw it run swiftly down the stairs and hide behind the same landscape painting where I first learned of its presence in my cabin. I moved nonchalantly to the steps, making my way upstairs, where I carefully positioned myself with string in hand, in full view of the box.

My heart was pounding as I watched the weasel inch closer to the box. First, it poked its head under the box to check out the meat. Then it moved into the box, with the exception of its tail. I was uncertain about what to do. I wanted to pull the string but was afraid the position of its tail would weaken the effectiveness of the trap. While I was briefly contemplating the situation, the weasel suddenly moved so that its tail was also inside the box. I pulled the string. The box fell faster than I had expected. The weasel was trapped!

What a commotion it made, scrambling around, hitting the interior of the box from all angles. I didn't waste any time; I quickly picked up the masonite, making certain to hold the box firmly in place, and ran to my car. The car door was open—I had prepared well in advance. I placed the box on the passenger side of the front seat, closed the door, and hastily drove away from my cabin.

It was a spectacle to behold. I was driving like a mad man, one hand on the wheel, the other hand on the box. I drove for about 8 mi., to a location well away from houses. I jumped out, placed the box on the frozen snow and promptly lifted it away from the masonite. Well, you should have seen that weasel tear off! It sped across the snow, heading directly for the trees, about 40 ft. away. It was the last time I saw it, except for the enduring visual image I have of it scampering over the snow.

II

For the remainder of the winter, Spirit lived a calm, peaceful life. In the spring, I placed him outside in his large living area as I had done for the last four years. He died in the latter part of the summer in 1994, in our fifth year together. Exactly how he died remains a mystery. I believe he was attacked by an animal such as a weasel. There was a hole in the chicken wire of his living space, which was large enough for a weasel to enter, when I searched his enclosure. All I found of Spirit were some feathers and his beak. I was devastated by my own loss and by what I imagined had been a violent death for Spirit. His life had already involved so much pain, it seemed unfair that his death should be violent, too. He had brought to my life much learning, sharing, laughter, and love. In my grief, I found myself searching for the best way to cope with his death.

III

Where there was much laughter, now there are many tears; those words that expressed my experience after I lost Spirit echoed in my mind as I walked the old dirt mining road towards Caribou Lake. Some time had passed since Spirit's death. I remember there was a chill in the air that early fall evening, and as I walked, the light from the sun reflected a soft glow from my red checked wool jacket.

In my left hand I carried a black hard-cover diary that contained written reflections about the five years Spirit and I had shared together. I stopped to listen to the sound of poplar leaves all about me. I opened my diary and read, "Notes on a Crow Named Spirit." I looked around, hearing the faint call of a crow. I tucked the diary in my knapsack and continued on my way.

A cool breeze floated over Caribou Lake. I noted

the impressions of old deer tracks in the soft mud. My eyes savoured every feature of this place. There was a special beauty that evening in the way the water and autumn leaves harmonized in colour and tranquility. It proved a good place to visit for the solitude I was seeking.

I went there to confront the many memories I had of Spirit, which were so vivid in my mind, and to silently express my deep gratitude for having known him. I also went there to reflect on my inadequacies, to reflect on whether I could have done more for Spirit. Perhaps I could have taken him flying more often, or I could have spent more time with him. As I stood on the shoreline, I felt acutely the difficulty of balancing everyday activities with the responsibility and need to share quality time with friends, human or otherwise.

I watched the sun pass behind a gnarled and weather-beaten pine, leaving its vestiges of light to dance over the waters of Caribou Lake. Soon dusk pressed heavily upon me, and daylight gave way to twilight, creating a magical mood, so characteristic of this place. I pulled the diary from my knapsack again. Slowly flipping its pages, I remembered the many experiences, which seemed to have happened just yesterday. A floodgate opened, and I glimpsed years of memories in a few moments; I felt a heavy sadness, accentuated by the sombreness of the landscape.

Suddenly, I exploded, screaming angry words at the animal that had destroyed Spirit's life! My words seemed to reverberate further and further, as if carrying my wrath away. In my mind's eye, I saw Spirit, I heard his greeting call, I felt the softness of his head in the palm of my hand, and I followed his flight to the soft pine needles on the forest floor. It all came in a rush, then I was left with the stillness of dusk. I sat, finally relaxed. From this stillness came a deep peace, and an understanding filled my heart, washing over the anger I had felt moments ago. I was humbled, emptied. In an instant the deep, peaceful stillness vanished, leaving only the darkness over Caribou Lake.

My experience at Caribou Lake was like a cleansing, a purging of much anger and sadness. It made a difference in my life. I was better able to accept Spirit's death, to feel that his life was truly for a purpose, and that we both played our parts well. We helped each other, in small ways, on the overall evolutionary path.

BIBLIOGRAPHY

Allen, Francis Henry. "The Aesthetic Sense in Birds as Illustrated by the Crow." *The Auk*. vol. 34, 1923. 112-113.

Angell, Tony. *Ravens, Crows, Magpies and Jays*. Seattle and London: University of Washington Press, 1978.

Bent, Arthur C. *Life Histories of North American Jays Crows, and Titmice*. New York: Dover Publications Inc., 1946.

Flicken, Millicent. "Avian Play." *The Auk*. vol. 94, 1977. 573-582.

Leahy, Christopher. *The Birdwatcher's Companion*. New York: Hill and Wang, a Division of Farrar, Strauss and Giroux; published simultaneously in Toronto: McGraw-Hill Ryerson Ltd., 1982.

Malefijt, Annemarie de Waal. *Religion and Culture: An Introduction to Anthropology of Religion*. New York: The Macmillan Company, 1968.

Pepperberg, Irene M. "A Communicative Approach to Animal Cognition: A Study of Conceptual Abilities of an African Gray Parrot." *Cognitive Ethology: The Minds of Other Animals*. Carolyn A. Ristau, ed., Hillsdale, NJ: Lawrence Erlbaum Associates, 1991. 159-160.

Pierce, Fred. "A Crow that Nearly Looped the Loop." *Wilson Bulletin*. vol. 34, 1923. 115.

Porter, J. P. "Intelligence and Imitation in Birds: A Criterion of Imitation." *American Journal of Psychology.* vol. 21, 1910. 1-71.

Powell, Robert W. "Operant Conditioning in the Common Crow (*Corvus brachyrhynchos*)." *The Auk*. vol. 89, 1972. 738-742.

Savage, Candace. *Bird Brains: The Intelligence of Crows, Ravens, Magpies and Jays*. Vancouver: Greystone Books, Douglas and McIntyre. 1995.

Sidney, Angela, Kitty Smith, and Rachel Dawson, Elders. *My Stories Are My Wealth*. As told to Julie Cruikshank for the Council of Yukon Indians. Whitehorse: Willow Printers. 1977.

The Holy Bible. (King James version). Nashville: Thomas Nelson Publishers, 1988.